Novel Materials for Sound-Absorbing Applications

Novel Materials for Sound-Absorbing Applications

Guest Editor

Eulalia Gliścińska

MDPI

Basel • Beijing • Wuhan • Barcelona • Belgrade • Novi Sad • Cluj • Manchester

Guest Editor
Eulalia Gliścińska
Faculty of Textiles and Design
Lodz University of
Technology
Lodz
Poland

Editorial Office
MDPI AG
Grosspeteranlage 5
4052 Basel, Switzerland

This is a reprint of the Special Issue, published open access by the journal *Materials* (ISSN 1996-1944), freely accessible at: https://www.mdpi.com/journal/materials/special_issues/9I9MA85U7A.

For citation purposes, cite each article independently as indicated on the article page online and as indicated below:

Lastname, A.A.; Lastname, B.B. Article Title. *Journal Name* **Year**, *Volume Number*, Page Range.

ISBN 978-3-7258-7158-2 (Hbk)
ISBN 978-3-7258-7159-9 (PDF)
https://doi.org/10.3390/books978-3-7258-7159-9

Contents

materials

Editorial

Novel Materials for Sound-Absorbing Applications

Eulalia Gliścińska

Textile Institute, Faculty of Textiles and Design, Lodz University of Technology, Zeromskiego Street, No. 116, 90-924 Lodz, Poland; klata@p.lodz.pl

Various places and situations are increasingly cause discomfort to humans due to the presence of disturbing sounds. The frequency of these sounds, depending on their origin, can widely vary, from low to high and from narrow to wide ranges [1]. Similarly, various possibilities and requirements exist for using materials to dampen unwanted sounds. Considering these various factors, new sound-absorbing materials with the required characteristics are being developed [2–10]. This Special Issue (SI), "Novel Materials for Sound-Absorbing Applications," comprises eight publications that advance the field of novel materials for sound-absorbing applications. This Special Issues includes the latest results from experimental, theoretical, and simulation studies.

Efforts are currently directed toward reducing and managing material waste. When designing new materials, their ease of processing and reuse must be considered [11]. In line with the principles of a circular economy, new materials should be produced using recycled materials. Therefore, research has focused on developing novel sound-absorbing materials whose components are sourced from recycled materials [12–14]. Technologies for such materials need to be developed, and the relationship of their sound absorption properties with the structure and composition of the materials must be studied [5,6,15–17]. Recyclate with a high degree of fragmentation poses a particular challenge, which hinders the ability to control the structure of the sound-absorbing material [13]. Badida et al. [18] designed a test insert for a measurement system, enabling the study of granular layered structures and comparatively analyzed the acoustic properties of rubber granulate from recycled car tires used in loose and compressed form. The test results indicated that the transmission losses were higher for more compressed and thicker samples composed of finer granulate fractions. Broda et al. [19] proposed using wool waste from tanneries as a natural, environmentally friendly material to produce carpet backing, which is the most versatile material for reducing noise in an enclosed space. This nonwoven backing offers an acoustic alternative to polyester nonwoven backing and helps reduce plastic pollution. Łach et al. [20] analyzed the influence of particle size and thermal treatment of diatomite on the use of diatomite as a component enhancing the sound absorption of composites. The authors descried the characteristics of diatomite, a mineral material obtained from Polish deposits, and a method for producing a green composite from natural components: flax fibers as reinforcement, polylactide fibers as the matrix material, and diatomite powder. The findings showed that sound absorption was increased without needing to calcinate diatomite and that the finest diatomite fraction should be used.

Designing sound-absorbing materials for reducing low-frequency or broadband sound remains a considerable challenge for engineers [8,13,21–23]. The presented studies focused on developing novel acoustic absorber structures, including those with a controllable resonance frequency and a tunable acoustic absorption bandwidth, demonstrating progress in the design of tunable acoustic absorbers. Zhang et al. [24] presented a novel solution for reducing broadband noise. They proposed a membrane metamaterial that was shunted

 https://doi.org/10.3390/ma19040740

using digitally controlled piezoelectric patches, which ensured high sound absorption at assumed programmable frequencies. Wang et al. [25] proposed replacing the micro-gap with a bent slit in a thin, 3.1 cm thick acoustic absorber, without changing the external dimensions. This replacement significantly increased low-frequency sound absorption and resulted in a broadened frequency band. This absorber effectively achieved adjustable resonance absorption and adjustable acoustic absorption bandwidth.

The development of new sound-absorbing materials with complex structures, such as foam, powder, honeycomb, or hybrids, necessitates the development of innovative tools for designing materials and methodologies for estimating and determining sound-absorbing properties [7,26–33]. In a theoretical analysis of the acoustic properties of lightweight powders, Sakamoto et al. [34] modeled a powder layer as a longitudinally oscillating one-dimensional beam and derived a transfer matrix. The proposed model requires only three experimentally determined parameters instead of nine, as in the commonly used Biot model [35]. Gai et al. [36] constructed an improved version of the honeycomb layered structure, a Helmholtz-type acoustic metamaterial, where each honeycomb is replaced with a small Helmholtz resonator. The authors developed theoretical and finite element models for this structure. The experimental studies confirmed that this solution ensures the material is light in weight and provides improved acoustic performance. Foam materials have structures that are geometrically difficult to model because of their irregularity. Sakamoto et al. [37] developed a method for estimating the sound absorption of foam materials with the membrane removed using computed tomography images.

The timely management of post-consumer and post-industrial waste, primarily from fibrous materials (e.g., carpets, disposable products, clothing, and technical products), paper, plastic, rubber, and natural waste, is urgently needed [13,14,16,38–40]. Future research should focus on developing new environmentally friendly sound-absorbing materials, i.e., recycled and/or recyclable materials [41]. This often requires the use of waste materials with varying degrees of fragmentation and that has various forms. Future work should therefore address the impact of the structure of sound-absorbing material containing shredded waste on the sound absorption performance. Despite the many existing solutions, enabling individuals to isolate from noise indoors or outdoors remains a major challenge. Future research should focus on the structural development of new sound-absorbing materials, i.e., those that meet indoor safety requirements or are suitable for outdoor use in air or water. These materials should have certain properties, e.g., constituting an additional thermal insulation or electromagnetic radiation barrier; being thin, i.e., less than 1 cm, but with high mechanical strength and resistance to various external factors; absorbing high amounts of sound across the widest possible frequency range; and reusing acoustic energy [16,42–45].

Conflicts of Interest: The author declares no conflicts of interest.

References

1. Sekar, V.; Cantwell, W.J.; Liao, K.; Berton, B.; Jacquart, P.-M.; Abu Al-Rub, R.K. Additively manufactured metamaterials for acoustic absorption: A review. *Virtual Phys. Prototyp.* **2024**, *19*, e2435562. [CrossRef]
2. Yang, M.; Sheng, P. Sound absorption structures: From porous media to acoustic metamaterials. *Annu. Rev. Mater. Res.* **2017**, *47*, 83–114. [CrossRef]
3. Glišcińska, E.; Michalak, M. Composite acoustic system for low-frequency sound absorption. *Sci. Rep.* **2025**, *15*, 4790. [CrossRef]
4. Glišcińska, E.; Michalak, M.; Michalak, A. Sound-absorbing thermoplastic composite with Helmholtz resonance. *Appl. Sci.* **2025**, *15*, 11349. [CrossRef]
5. Hajizadeh, R.; Montazeri, A.; Esnaasharieh, M.; Mosayebian, M.; Beheshti, M.H.; Hosseini, H. Investigation of the effect of using recycled polypropylene waste on the sound absorption coefficient of micro-perforated panels as a green absorber. *Int. J. Environ. Sci. Technol.* **2025**, *22*, 10599–10612. [CrossRef]

6. Yun, B.Y.; Cho, H.M.; Kim, Y.U.; Lee, S.C.; Berardi, U.; Kim, S. Circular reutilization of coffee waste for sound absorbing panels: A perspective on material recycling. *Environ. Res.* **2020**, *184*, 109281. [CrossRef]
7. Huang, S.; Li, Y.; Zhu, J.; Tsai, D.O. Sound-absorbing materials. *Phys. Rev. Appl.* **2023**, *20*, 010501. [CrossRef]
8. Sujon, M.A.S.; Islam, A.; Nadimpalli, V.K. Damping and sound absorption properties of polymer matrix composites: A review. *Polym. Test.* **2021**, *104*, 107388. [CrossRef]
9. Liang, M.; Wu, H.; Liu, J.; Shen, Y.; Wu, G. Improved sound absorption performance of synthetic fiber materials for industrial noise reduction: A review. *J. Porous Mater.* **2022**, *29*, 869–892. [CrossRef]
10. Kalauni, K.; Pawar, S.J. A review on the taxonomy, factors associated with sound absorption and theoretical modeling of porous sound absorbing materials. *J. Porous Mater.* **2019**, *26*, 1795–1819. [CrossRef]
11. Yang, T.; Hu, L.; Xiong, X.; Petru, M.; Mishra, R.; Militký, J. Sound absorption properties of natural fibers: A review. *Sustainability* **2020**, *12*, 8477. [CrossRef]
12. Selvaraj, V.K.; Subramanian, J.; Dutt, S.S.; Annamalai, K.; Natarajan, E.; Kumaresan, S. A comparative study on acoustical properties using waste recycled porous materials for environmental sustainability. *Sci. Rep.* **2025**, *15*, 26460. [CrossRef]
13. Zeleznik, A.; Curovic, L.; Prezelj, J. Sustainable sound absorption using shredded plastic particles: Adjusting low-frequency acoustic performance with particle size. *J. Low Freq. Noise Vib. Act. Control* **2025**, *44*, 1139–1150. [CrossRef]
14. Kang, C.-W.; Choi, B.-S.; Hasegawa, M.; Oh, J.S.; Kolya, H. Innovative approaches for recycling office waste paper with potential for sound-absorbing wall panels. *J. Mater. Cycles Waste Manag.* **2025**, *27*, 2444–2453. [CrossRef]
15. Astrauskas, T.; Grubliauskas, R.; Janusevicius, T. Optimization of sound-absorbing and insulating structures with 3D printed recycled plastic and tyre rubber using the TOPSIS approach. *J. Vib. Control* **2024**, *30*, 1772–1782. [CrossRef]
16. Singh, E.A.; Dubey, N.; Punia, S. Eco-friendly noise-absorbing panels from waste materials: A sustainable solution to control noise pollution. *Int. J. Multidiscip. Res.* **2024**, *6*, 240113034.
17. Özdil, N.; Özçelik Kayseri, G.; Süpüren Mengüç, G. *Investigation of Sound Absorption Characteristics of Textile Materials Produced from Recycled Fibers Waste in Textile and Leather Sectors*; IntechOpen: London, UK, 2020. [CrossRef]
18. Badida, M.; Andrejiova, M.; Pinosova, M.; Moravec, M. Acoustic analysis of soundproofing materials using recycled rubber from automobiles. *Materials* **2025**, *18*, 3144. [CrossRef]
19. Broda, J.; Kobiela-Mendrek, K.; Baczek, M.; Rom, M. Aacoustic performance of tufted carpets coupled with underlayment produced from tannery wool waste. *Materials* **2025**, *18*, 315. [CrossRef]
20. Łach, M.; Gliścińska, E.; Przybek, A.; Smoroń, K. The influence of diatomite on the sound absorption ability of composites. *Materials* **2024**, *17*, 4590. [CrossRef]
21. Chang, L.; Jiang, A.; Rao, M.; Ma, F.; Huang, H.; Zhu, Z.; Zhang, Y.; Wu, Y.; Li, B.; Hu, Y. Progress of low-frequency sound absorption research utilizing intelligent materials and acoustic metamaterials. *RSC Adv.* **2021**, *11*, 37784–37800. [CrossRef]
22. Mohammadi, M.; Ishak, M.R.; Sultan, M.T.H.; Zainudin, E.S. Comprehensive review of factors influencing the sound absorption properties of micro-perforated panel structures. *J. Vib. Eng. Technol.* **2025**, *13*, 319. [CrossRef]
23. Naimusin, A.; Kizinievic, O.; Astrauskas, T.; Janusevicius, T. Layered acoustic structures made of plastic microfibre pulp waste with metamaterial for low- and middle-frequency sound absorption applications. *J. Low Freq. Noise Vib. Act. Control* **2025**, *44*, 2242–2260. [CrossRef]
24. Zhang, X.; Nie, J.; He, J.; Lin, F.; Liu, Y. Digitally controlled piezoelectric metamaterial for low-frequency and high-efficiency sound absorption. *Materials* **2025**, *18*, 2102. [CrossRef]
25. Wang, H.; Ma, P.; Fan, X. A compact low-frequency acoustic perfect absorber constructed with a folded slit. *Materials* **2024**, *17*, 5992. [CrossRef]
26. Sagartzazu, X.; Hervella, L.; Pagalday, J.M. Review in sound absorbing materials. *Arch. Comput. Methods Eng.* **2007**, *15*, 311–342. [CrossRef]
27. Liu, R.; Zhang, Z.; Zheng, X. The modern numerical and experimental methods for the sound absorbing characteristics of dissipative sound absorbing materials: A review. *Materials* **2025**, *18*, 5353. [CrossRef] [PubMed]
28. Bobrovnitskii, Y.I.; Tomilina, T.M. Sound absorption and metamaterials: A review. *Acoust. Phys.* **2018**, *64*, 519–526. [CrossRef]
29. Cao, L.; Fu, Q.; Si, Y.; Ding, B.; Yu, J. Porous materials for sound absorption. *Compos. Commun.* **2018**, *10*, 25–35. [CrossRef]
30. Dunne, R.; Desai, D.; Sadiku, R. A review of the factors that influence sound absorption and the available empirical models for fibrous materials. *Acoust. Aust.* **2017**, *45*, 453–469. [CrossRef]
31. Du, Y.; Wu, W.; Chen, W.; Lin, Y.; Chi, Q. Control the structure to optimize the performance of sound absorption of acoustic metamaterial: A review. *AIP Adv.* **2021**, *11*, 060701. [CrossRef]
32. Otaru, A.J. Review on the acoustical properties and characterisation methods of sound absorbing porous structures: A focus on microcellular structures made by a replication casting method. *Met. Mater. Int.* **2020**, *26*, 915–932. [CrossRef]
33. Babacan, O. An artificial intelligence approach to evaluating the sound absorption performance of materials: A systematic review. *Düzce Univ. J. Sci. Technol.* **2025**, *13*, 718–734. [CrossRef]

https://doi.org/10.3390/ma19040740

34. Sakamoto, S.; Kawakami, Y.; Soeta, H.; Kubo, Y. Theoretical analysis of low-frequency sound absorption owing to the vibration of lightweight powder using a 1D beam model. *Materials* **2025**, *18*, 2611. [CrossRef]
35. Biot, M.A. Theory of propagation of elastic waves in a fluid-saturated porous solid. I. Low-frequency range. *J. Acoust. Soc. Am.* **1956**, *28*, 168–178. [CrossRef]
36. Gai, X.-L.; Li, X.-H.; Guan, X.-W.; Xing, T.; Cai, Z.-N.; Hu, W.-C. Study on acoustic properties of Helmholtz-type honeycomb sandwich acoustic metamaterials. *Materials* **2025**, *18*, 1600. [CrossRef] [PubMed]
37. Sakamoto, S.; Satoh, T.; Tanabe, K.; Maruyama, K.; Himori, Y. Theoretical estimation based on CT images and experiments on the sound absorption coefficient of foam materials with removed membrane. *Materials* **2025**, *18*, 846. [CrossRef] [PubMed]
38. Santhanam, S.; Bharani, M.; Temesgen, S.; Atalie, D.; Ashagre, G. Recycling of cotton and polyester fibers to produce nonwoven fabric for functional sound absorption material. *J. Nat. Fibers* **2019**, *16*, 300–306. [CrossRef]
39. Gaj, P.; Kopania, J.; Wójciak, K.; Bogusławski, G. Assessment of sound absorbing properties of composite made of recycling materials. *Vib. Phys. Syst.* **2019**, *30*, 2019109.
40. Turkiewicz, J.; Sikora, J. Sound absorbing materials from recycled rubber products. *Mech. Control* **2013**, *32*, 117–121. [CrossRef]
41. Gomeza, T.S.; Zuluagaa, S.; Jimeneza, M.; de la Pridac, D.; Pedrerod, A.; Navacerradad, M.A.; Restrepo-Osorioa, A.; Fernandez-Morales, P. Sound absorption behavior of repurposed waste fibers: Effects of fiber size, density, and binder concentration. *Appl. Acoust.* **2023**, *202*, 109174. [CrossRef]
42. Tao, Y.; Zhang, M.H.; Peijs, T. Recent progress in acoustic materials and noise control strategies—A review. *Appl. Mater. Today* **2021**, *24*, 101141. [CrossRef]
43. Fu, Y.U.; Kabir, I.I.; Yeoh, G.H.; Peng, Z.X. A review on polymer-based materials for underwater sound absorption. *Polym. Test.* **2021**, *96*, 107115. [CrossRef]
44. Saha, K.; Sonawwanay, P.D.; Patel, V.; Solanki, M. A review on innovative measures for improving the sound absorption of acoustic materials for broader frequency ranges. *Noise Vib. Worldw.* **2025**. [CrossRef]
45. Zhang, X.; Qu, Z.; Wang, H. Engineering acoustic metamaterials for sound absorption: From uniform to gradient structures. *iScience* **2020**, *23*, 101110. [CrossRef] [PubMed]

https://doi.org/10.3390/ma19040740

materials

MDPI

Article

Acoustic Analysis of Soundproofing Materials Using Recycled Rubber from Automobiles

Miroslav Badida [1], Miriam Andrejiova [2], Miriama Pinosova [3] and Marek Moravec [1,*]

[1] Department of Business Management and Economics, Technical University of Košice, 042 00 Košice, Slovakia; miroslav.badida@tuke.sk

[2] Department of Applied Mathematics and Informatics, Technical University of Košice, 042 00 Košice, Slovakia; miriam.andrejiova@tuke.sk

[3] Department of Quality, Safety, and Environment, Technical University of Košice, 042 00 Košice, Slovakia; miriama.pinosova@tuke.sk

* Correspondence: marek.moravec@tuke.sk

Abstract

This article provides a comprehensive analysis of the acoustic properties of recycled rubber crumb, examined in two forms—loose granular and compacted specimens. The aim was to compare their acoustic properties depending on the size of the fraction, the thickness of the sample, and the degree of compaction, with measurements performed using a model BSWA SW433 impedance tube in the frequency band 100–2500 Hz. Experimental samples of recycled rubber crumb were prepared with various thicknesses (2, 4.5, and 7 cm) and of various fractions (0–4 mm), and the granular samples were compacted under a pressure of 250–750 kPa. The results showed that the highest transmission loss (TL) is achieved by fine fractions at higher pressure and with greater sample thickness; Fraction 1 (below 1 mm) at a pressure of 750 kPa and a thickness of 7 cm had the best acoustic properties. Through regression analysis, mathematical models of the dependence of transmission loss on the monitored parameters for all types of samples (granular/compacted) were created. The regression analysis confirmed that the thickness, pressure, and size of the fraction significantly affect the acoustic properties of the material. Recycled rubber crumb therefore represents an efficient and environmentally sustainable alternative to traditional insulation materials, and optimizing its parameters enables a wide range of practical acoustic applications in construction, transport infrastructure, and manufacturing industries.

Keywords: recycled rubber crumb; acoustic properties; transmission loss; measurement; impedance tube

1. Introduction

The acoustic properties of materials play a critical role in developing and implementing noise control measures across various industries. Analyzing these properties is essential for effective noise mitigation [1,2]. Optimizing the acoustic environment depends on selecting suitable construction and sound-absorbing materials, as well as the methods used for their application. Environmental sustainability is also a key consideration in designing acoustic materials, especially given the increasing consumption of raw materials, the growth of plastic waste in ecosystems, and global environmental pollution. Recycling rubber waste is one of the main challenges in end-of-life vehicle recycling. Due to their unique physical and chemical properties, recycled rubber materials can provide significant benefits in acoustic and thermal insulation, as well as vibration control, when used appropriately.

Desarnaulds et al. [3] described traditional insulation materials, which are primarily used in building acoustics because of their favorable environmental and acoustic properties. These materials fall into two main categories: natural materials, such as cotton, cellulose, or hemp, and recycled materials, including rubber, cork, glass, textiles, and other secondary raw materials. Natural materials have a low environmental footprint throughout their life cycle. Their main advantages are complete recyclability and biodegradability. Additionally, most of these materials do not require binders, as they can be compacted using a pressing device [4–6]. McGinnes et al. [7] highlighted the benefits of natural materials, such as straw, in the construction industry, emphasizing their availability, ease of use, and environmental friendliness. They found that the sound absorption of these materials is comparable to that of mineral wool or glass fibers. However, D'Alessandro et al. (2017) refuted this claim, showing that straw bale walls are weak sound insulators at low frequencies and do not meet building code requirements [8]. The authors [4] pointed out that natural and recycled materials offer cost and environmental advantages over synthetic materials. However, the acoustic properties of many natural materials remain insufficiently studied.

A new approach to the reuse of shredded recycled rubber was presented by the authors in a study [9], in which they highlighted the growing problem of the large number of worn-out tires generated each year. As part of their research, they developed a new material with potential applications in construction as both an acoustic and thermal insulation element. They also emphasized that the use of crushed recycled rubber in these areas is only sporadically addressed in the available literature, with the exception of a few publications, such as the studies by Sambucci et al. [10–12].

Akiwate et al. [13] investigated the acoustic properties of porous materials produced using 3D printing technology. Labašová and Ďuriša [14] evaluated the acoustic absorption coefficients of various single- and double-layer materials, such as cork, felt, and polyethylene. Sandu et al. [15] tested 17 different materials—including recycled options such as cinder, polystyrene, cork, balsa wood, and bird feathers—to identify those with the highest potential for noise reduction. The authors [16] investigated the sound insulation properties of recycled textile fiber composites, evaluating the acoustic properties using an impedance tube in the frequency range of 100–3200 Hz. Their results showed that absorption improved with increasing composite layer thickness. Research by the authors [17] on the acoustic properties of concrete, mortar, and stabilized soil revealed that these materials achieved the highest absorption at low frequencies (100–500 Hz). Concrete exhibited the highest sound absorption coefficient (0.94), followed by mortar (0.85) and stabilized earth (0.74). These findings highlight significant differences in the acoustic properties of individual materials, which are important considerations for their application in the construction industry, particularly regarding sound insulation.

The impedance tube has been used in experimental studies by Suhanek M. et al., Pierre J. et al., Koruk H., Tan W.H. et al., and Neto B.N. et al. [18–22]. These studies demonstrate the versatility of impedance tube measurements for evaluating the acoustic properties of various recycled materials, highlighting their potential applications in sustainable building and noise reduction solutions.

The selection of recycled rubber granulate from automotive tires as the primary material for this study was motivated by environmental sustainability, as tire recycling represents a significant challenge in automotive waste management. The material is also readily available due to a well-developed recycling infrastructure and is cost-effective compared to traditional acoustic materials, which supports its broader application in industry. The acoustic potential of the granulate stems from its porous structure and the possibility to adjust its fraction, thickness, and compaction.

Our article presents a significant contribution to the scientific literature by presenting a comprehensive experimental analysis of the acoustic properties of recycled rubber crumb in various forms and configurations. This includes the development of a patented measuring device. We develop regression models to predict acoustic performance, compare the effectiveness of recycled rubber with that of commercial materials, and provide practical, environmentally conscious recommendations. Consequently, this study expands current knowledge on the use of recycled materials in acoustics, supplementing existing research with new data, predictive models, and innovative methodologies.

This study examines the acoustic properties of recycled rubber crumb in two forms: loose granular and compact specimens. The objective is to compare their sound absorption capabilities and evaluate their potential for soundproofing applications. Measurements were made using an impedance tube, and a cartridge was developed for granular materials, enabling extended measurement options for these materials. Experimental samples of recycled rubber crumb were prepared with varying thicknesses (2 cm, 4.5 cm, 7 cm) and fraction sizes ($\leq$1 mm, 1–2 mm, 2–3 mm, 3–4 mm). The samples were compacted under pressures of 250 kPa, 500 kPa, and 750 kPa using a filling device within test cartridges (inner diameter: 4.9 cm). For comparison, compact samples were cut from commercially available panels in the same thicknesses (2 cm, 4.5 cm, 7 cm) with a diameter of 5.9 cm. These panels are costly to produce, whereas recycled crumb offers a cost-effective and eco-friendly alternative. The research focused on analyzing how fraction size, thickness, and compaction pressure influence the transmission loss parameter (TL, in dB).

2. Materials and Methods

Acoustic materials, as noted earlier, are widely used in industrial and everyday environments to reduce noise and improve acoustic conditions. They are increasingly applied in homes, offices, and production facilities. In construction, these materials serve as inter-floor layers, partitions, or insulation boards to enhance sound insulation and mitigate impact noise. In transportation infrastructure, they line tunnels, airports, and platforms to dampen traffic noise. Industrial applications include anti-vibration membranes and insulation panels to reduce structural vibration and noise transmission. In healthcare settings, they improve acoustic conditions in clinics and operating rooms [23,24].

In the context of environmental sustainability and material recycling, rubber recyclate (crumb) represents a viable alternative for acoustic applications. Shredding waste rubber into granules (0.5–4 mm grain size) creates a heterogeneous porous structure with a sound absorption coefficient of $\alpha \geq 0.6$ in medium and high frequencies (above 1000 Hz) [25]. Pressing or binding the crumb with polymer adhesives produces compact samples that, due to their homogeneous composition and optimized thickness, provide consistent noise attenuation even at low frequencies [23]. These compact rubber modules achieve transmission loss (TL) levels comparable to conventional synthetic foams while offering superior resistance to mechanical wear and moisture [25–27]. Key parameters—such as rubber crumb grain size, binder quantity, and pressed sample geometry—can be tailored to specific application requirements, enabling broad use in construction, transportation, and industry. Modular production processes support customized solutions, ranging from thin acoustic inserts for partitions to robust noise barriers and flooring systems. Implementing rubber recyclate and its compact variants not only advances environmental sustainability through recycling but also delivers cost-effective, high-performance acoustic solutions across practical applications.

The obtained data were then statistically processed using several software tools, namely Minitab in version 18 (regression analysis), Python version 3.10.11 (using the Pandas version 2.2.3 and Seaborn libraries version 0.13.2), program R version 4.4.3 (Stat package), and OriginPro 2019b for the graphical processing of the results. Results include regression models and visualizations illustrating the dependence of transmission loss on the tested variables.

2.1. Development and Construction of Test Cartridges for the Experimental Evaluation of Acoustic Properties of Granular Materials

The development of test cartridges for measuring acoustic properties was necessary for a number of reasons, and it is one of the key elements of this research. Standard impedance tubes are primarily designed for testing compact samples and are therefore not suitable for directly evaluating loose granular materials. To overcome this limitation, we designed a new type of cartridge that allows for the safe and accurate placement of granular samples within the measurement system.

The main innovation of this cartridge is its versatility; it enables the testing of not only granular materials of various fractions but also pressed, compact samples. The cartridge's design allows for the creation of sandwich structures (e.g., granular material–air gap–compact layer–granulate), while the filling module (compression device) ensures the regulation of the density of the material (200–800 kg/m^3) at operating pressures up to 750 kPa. In this way, it is possible to accurately simulate real layering in structures and analyze the effect of compaction on the acoustic properties of materials such as glass wool, rubber crumb, or textile fibers. The test cartridge thereby not only serves as a measuring element but also functions as an integration element of the filling system.

The cartridge body is made of aluminum alloy (Al 6082-T6) with a wall thickness of 5 mm. The perforated screens are made from 1.4301 stainless steel, with 3 mm diameter holes spaced 2 mm apart, which prevents granular material from falling through. The inner diameter of the cartridge is 49 mm. The sample height is variable—20, 50, 75, 100, and 120 mm—enabling different thicknesses to be tested without the need to replace the entire cartridge. Figure 1 shows the cartridge design, featuring a threaded connection between the cartridge body and the two perforated sections (bottom and lid), which enables easy filling and emptying of the material to be tested.

The validation procedure included verifying the uniformity of the acoustic field using reference measurements with polyurethane foam for calibration (250 kg/m^3, $\alpha = 0.95 \pm 0.03$). This confirmed that the cartridge does not create unwanted reflections or distort the wave field in the measurement area. Sample compression was controlled by pressing the samples at 5 kPa, 10 kPa, and 20 kPa, while monitoring changes in both thickness and volumetric density of the material.

The main advantages include the following:

1. The proposed design allows for the measurement of both granular and compact samples without the need to modify the entire measurement setup.
2. The mechanism enables fast and precise adjustment of sample thickness with high repeatability, without interfering with the measurement equipment.
3. The integrated pressing and filling system with adjustable pressure enables regulation of the density (kg/m^3) of the material to be examined.

Because of these properties, it is possible to easily investigate the effect of thickness and pressure on transmission loss (TL, dB). Additionally, the modular design of the cartridge ensures that a single component can function as both a test and filling module.

The utility model of the cartridge was registered by the Industrial Property Office of the Slovak Republic under number 9662, which confirms the originality and practical usability of this solution.

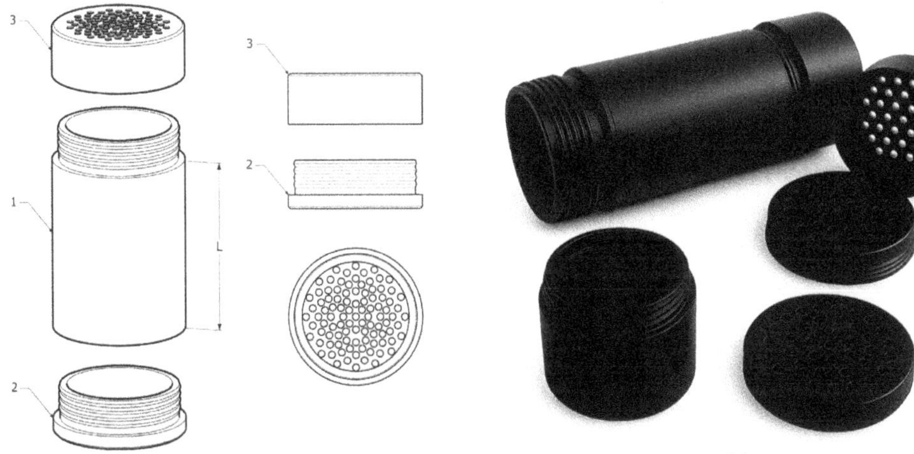

(a) (b)

Figure 1. Test cartridge with interchangeable perforated parts and threaded cap. (**a**) Technical drawing: 1—Cartridge body (intended for insertion of test materials), 2—Lower perforated bottom/sieve (retains the material inside the cartridge), 3—Upper perforated lid/sieve (and also the bottom prevents the spilling of material), L—length of the cartridge, which enables adaptation to different lengths of test samples; (**b**) 3D model of the test cartridge and its individual parts.

2.2. Semi-Automated Equipment for Filling and Compacting Granular Materials in Test Cartridges

A semi-automated device was developed for filling test cartridges with granular materials (Figure 2), significantly expanding the capabilities for measuring the acoustic properties of recycled granular materials from end-of-life vehicles. This equipment allows for the application of various pressures to the granular material, enabling controlled sample compaction. As a result, it is possible to analyze the influence of compaction and thickness on the acoustic properties of the tested samples, specifically transmission loss.

The device is constructed of high-quality aluminum, while the protective cover is made of a robust steel sheet, which ensures high durability and a long service life. The overall dimensions of the device are 410 mm × 754 mm × 414 mm and its weight is 15 kg, which enables easy handling even in laboratory conditions. When measuring the noise at a distance of 30 cm from the device, a value of 50 dB was recorded, which means that the device works relatively quietly and does not disturb the surroundings. The function of the device is to press the granular material inside the cartridge. The procedure is as follows:

1. Preparation: The required amount of recycled material is poured into the cartridge, ensuring that it is evenly distributed throughout the cartridge volume. The cartridge is then inserted into the centering ring, which guarantees accurate and repeatable cartridge placement in the device, minimizing the risk of improper pressing and ensuring consistency of results in further sample processing.
2. Pressing: Using a linear piston driven by an electric motor, the material is compressed to the desired pressure. The compression force is displayed on the control panel, and the device allows pressure settings from 250 to 750 kPa in 50 kPa increments. This range provides precise and repeatable compaction tailored to the specific application requirements.
3. Closure: After compaction, the lid is securely screwed onto the cartridge, maintaining the stability and integrity of the sample during handling and subsequent measurement. This step is important to maintain the desired degree of compaction and to

prevent undesirable changes in the properties of the sample before it is inserted into the impedance tube.

4. Preparation for measurement: The compacted sample is then inserted into an impedance tube for accurate measurement of its acoustic properties, specifically transmission loss (TL, dB). Measurements follow the STN EN ISO 10534-2 standard, typically using the transfer function method, which employs two or more microphones to analyze the standing wave within the tube. The test cartridge prevents material spillage and ensures measurement repeatability. Results are processed using VA-Lab4 software, providing data on the acoustic behavior of the recycled material across the frequency range of 100 Hz to 2.5 kHz.

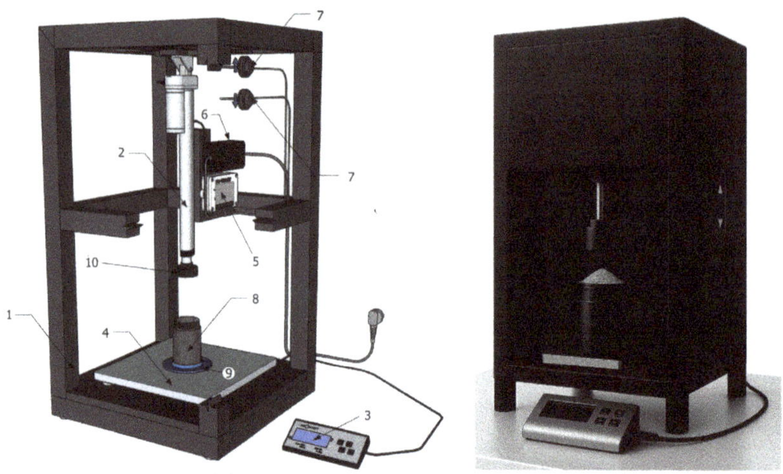

(**a**) (**b**)

Figure 2. Laboratory filling and compaction apparatus with digital control. (**a**) Technical drawing: 1—Equipment frame, 2—Linear electric piston, 3—Control panel, 4—Integrated scale, 5—Control unit, 6—Power supply, 7—Vertical movement buttons (up/down), 8—Test cartridge, 9—Granular material holder, 10—Pressure mechanism; (**b**) 3D model of the device and its individual parts.

The device is designed for efficient and repeatable pressing granular or crushed/ground materials, which is essential for accurate acoustic and material analyses. Its robust aluminum and sheet steel construction ensures a long service life and reliability even with frequent use. With its intuitive operation, the ability to precisely adjust the pressure over a wide range and the automated electric drive, the device is ideal for laboratories, research facilities and industrial operations where high accuracy and repeatability of sample preparation are required.

The semi-automated equipment for filling and compacting granular materials in test cartridges has received a utility model approved by the Industrial Property Office of the Slovak Republic under registration number 9665. This entry officially confirms the original design and practical usability of the device in laboratory and industrial conditions.

2.3. Characteristics of the Material Samples Tested

The recycled rubber crumb in various fractions was supplied for research purposes by a company specializing in the processing of waste tires and the production of rubber granulate. This material is widely used in the construction industry, particularly as an additive in asphalt mixtures at proportions of 15–22%, enhancing the flexibility and durability of road surfaces. Additionally, recycled rubber crumb is commonly used as infill for sports fields, contributing to shock absorption and increased surface safety. The supplier is one of the largest producers of rubber granulate in Slovakia, with operations encompassing comprehensive waste management, including recycling and recovery of secondary raw materials.

Granular samples were obtained from waste tires by means of a mechanical crushing process. The obtained rubber crumb was sieved through four circular sieves under laboratory conditions to obtain the required size fractions (up to 1 mm, 1–2 mm, 2–3 mm, 3–4 mm) for the experimental samples. This procedure ensured accurate control of the granulometry of the material, which was important for evaluating the acoustic properties of individual fractions. The compact panels were made of granulate fraction 0–4 mm using the CONIPUR 372/1 polyurethane binder (CONICA AG, Schaffhausen, Switzerland), which provides high mechanical resistance and stability at temperatures up to 150 °C. The compaction process took place at a temperature of 80–100 °C and a pressure of 2–3 MPa for 30–60 min, which ensured a homogeneous structure without internal defects. Experimental samples of granular crumb were prepared in three thicknesses (2 cm, 4.5 cm, 7 cm), while compact samples were waterjet cut to equal thicknesses with a diameter of 5.9 cm, ensuring dimensional accuracy and material integrity without thermally affecting the structure. The samples were made in two basic versions, namely loose granular (p = 0 kPa) and compacted under pressure (250 kPa, 500 kPa, 750 kPa) in test cartridges with an inner diameter (4.9 cm) using a device for filling them.

This study aims to compare the acoustic parameters between granular and compact forms of recycled rubber, using an impedance tube measurement methodology. The choice of materials was influenced by the pursuit of environmental sustainability, the easy availability of raw materials, and the expected acoustic properties, which are also confirmed by relevant scientific studies [25,28–31]. The properties of individual tested samples are listed in Table 1.

Table 1 shows the average values of the specific mass (ρ, kg/m^3) of recycled rubber granulate, depending on the size fraction, which were determined by laboratory measurement of loose granular material without external compaction pressure. The relative deviations of measurements for granular materials do not exceed 5%, which corresponds to the variability of experimental conditions, such as the heterogeneity of the sample and the accuracy of the measuring instruments used.

2.4. Measurement of Transmission Loss (TL) Using an Impedance Tube

Several measurement methods have been developed to determine transmission loss (TL), with one of the most widely used being impedance tube measurement. The impedance tube has an advantage over other measurement methods due to its compactness and the ability to obtain results quickly. For example, compared to an echo chamber, it requires significantly smaller samples of the material to be tested.

In this study, transmission loss of recycled rubber materials was measured using a BSWA SW433 impedance tube (BSWA Technology Co., Ltd., Beijing, China) in a four-microphone configuration, with a length of 500 mm, an inner diameter of 60 mm, and a frequency range of 100–2500 Hz (see Figure 3). The measurement setup also included a 4-channel MC3242 analyzer (BSWA Technology Co., Ltd., Beijing, China) (0–20 kHz), a PA50 amplifier (BSWA Technology Co., Ltd., Beijing, China) (50 W) to power the tube's speaker, a computer with VA-Lab4 software version 6.0, the necessary wiring, and the impedance tube itself. Transmission loss measurement was performed using the transformation function method in accordance with STN EN ISO 10534-2 (730537) [32,33]. Five repeated measurements were taken for each sample to ensure statistical reliability. To assess the impact of sample thickness, particle size fraction (rubber granules versus compact samples), and compaction pressure on transmission loss (TL, dB), experimental samples were prepared with standardized parameters, as described in detail in Section 2.3.

Table 1. List and sample parameters of recycled rubber materials.

Material	Structure	Fraction (mm)	Sample Views Macro/Micro Views	Bulk Density (kg/m^3)
Recycled rubber panel	compact	0–4		849
Recycled rubber crumb	granular	≥1	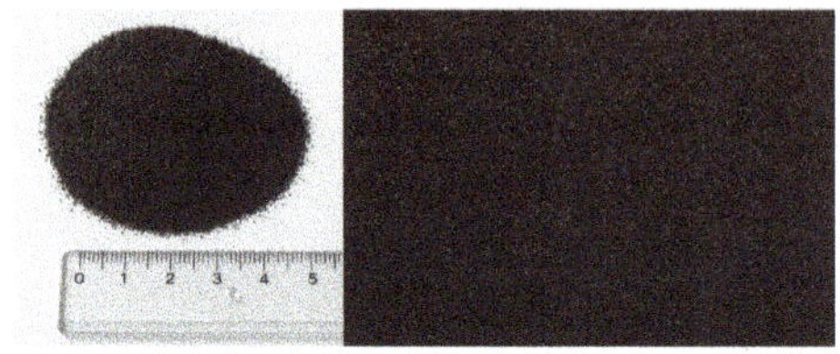	377
		1–2	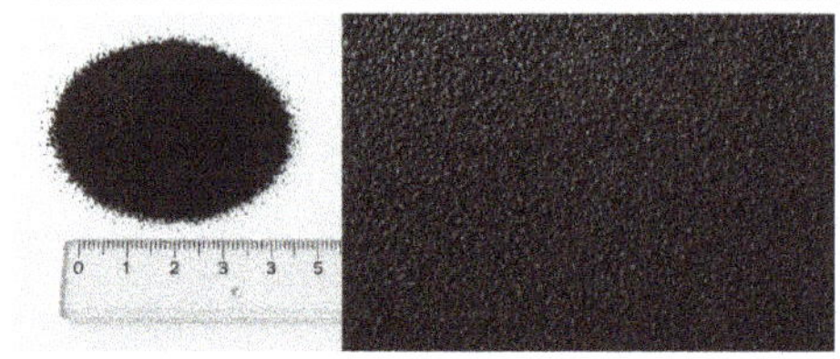	458
		2–3	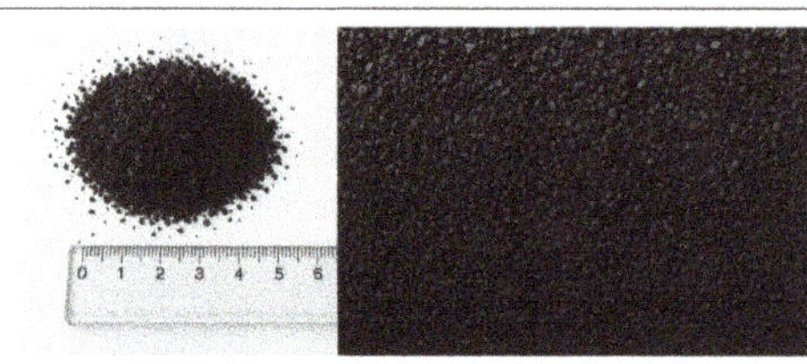	472
		3–4	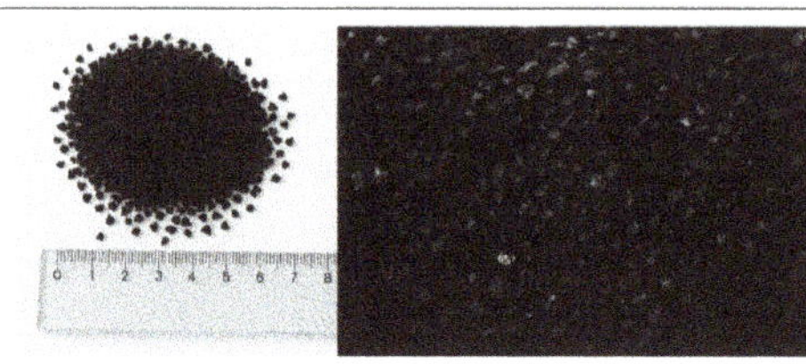	500

Key parameters of sample preparation included the following:

1. Compaction of samples using calibrated equipment with precisely set pressure (for more details, see Section 2.2).
2. Sorting rubber crumb particles by sieving through sieves with standardized holes.
3. Stabilization of samples in an air-conditioned environment prior to measurement.

This procedure ensured a high degree of reproducibility and relevance of the experimental data obtained.

Temperature (22.5 ± 1 °C) and relative humidity (48 ± 5% RH) values were monitored throughout the experiment. An atmospheric pressure of 986.07 hPa indicated a stable laboratory environment suitable for reproducible measurements.

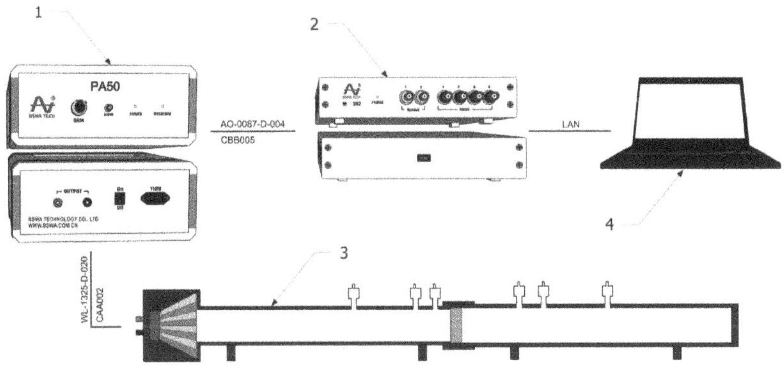

Figure 3. Connection diagram for measuring equipment [32]. Explanatory note: 1—PA50 power amplifier (50 W) for driving the speaker in the impedance tube; 2—MC3242 data acquisition analyzer with 4 ICP input channels and 2 output channels (0~20 kHz); 3—Impedance tube BSWA SW433; 4—PC with Va-Lab4 software.

Transmission loss was determined as a decibel (dB) value that expresses the ratio between the intensity of the sound falling on the front of the sample and the intensity of the sound transmitted through the test material. This parameter enabled us to assess how effectively the material dampens sound—the higher the transmission loss value, the better the damping properties of the material. The measurement was performed in a BSWA SW433 impedance tube, where we transmitted a broadband stationary signal PA50 (50 W) to the material sample. Using an array of microphones, we recorded the incident and reflected component of the sound on the front of the sample, and also the transmitted and reflected component on the back of the sample. This approach enabled precise determination of the proportion of sound energy transmitted, reflected, or absorbed by the material.

During the measurement, we monitored the influence of several factors: the frequency of mechanical oscillation, the thickness of the material, its internal structure (thickening under pressure—reducing porosity and increasing density), and the temperature of the environment.

We calculated the transmission loss (TL) according to the following equation:

$$TL = 10 \log_{10} \left| \frac{W_i}{W_t} \right|, \tag{1}$$

where W_i is the energy of the incident sound wave that comes towards the test sample, W_t is the energy of the transmitted sound wave that departs from the test sample.

2.5. Methods of Data Analysis and Processing

2.5.1. Transfer Matrix Method

For the assessment and comparison of the determined transmission loss, terms describing the acoustic properties of the examined samples will be defined: transmission loss (TL_f), transmission coefficient (τ_f), average transmission coefficient (τ_{avg}), relative transmission ratio (RTR_f), average transmission loss (TL_{avg}).

Transmission loss (TL, dB) is defined as the logarithm of the ratio of the incident and transmitted sound wave (power or intensity); see Equation (1).

Equation (2) for transmission loss, which is equivalently expressed by transmission coefficient (τ_f) at frequency f, can be written as follows:

$$TL_f = -10\log_{10}\left(\tau_f\right), \tag{2}$$

where τ_f is the proportion of the transmission power.

To correctly add our average decibel (dB) values, each value must first be converted from dB to linear values. This is because a decibel is a logarithmic unit expressing the ratio of two values (such as power or intensity) and therefore cannot be treated directly as a regular numerical value.

From each measured value of the transmission loss at frequency f, we calculate the linear power transmission coefficient according to Equation (3):

$$\tau_f = 10^{-\frac{TL_f}{10}}, \tag{3}$$

where TL_f is the respective transmission loss at frequency f.

The power transmission coefficient expressed as a percentage $\tau_f \times 100\%$ indicates the ratio of the sound power transmitted by the material or structure to the total incident power across the entire frequency band (in our case 100–2500 Hz). A higher transmission percentage means better transmission through the material and less power loss.

Interpretation of the value of τ_f:

1. If $\tau_f \times 100\% \rightarrow 100\%$, the output power equals the input power, meaning there is no attenuation (0 dB attenuation).
2. If $\tau_f \times 100\ \% \rightarrow 50\%$, about 50% of the input power is transmitted, which corresponds to a transmission loss of approximately 3 dB.
3. If $\tau_f \times 100\% \rightarrow 10\%$, about 10% of the input power is transmitted, which corresponds to a transmission loss of approximately 10 dB.
4. If $\tau_f \times 100\% \rightarrow 0\%$, almost no input power is transmitted, indicating nearly complete attenuation.

The value of $1/\tau_f$ expresses the ratio of input power to output power; in other words, it indicates how much greater the input power is compared to the transmitted (output) power. Next, we calculate the energy (linear) average of the transmission coefficient τ_{avg} for the frequency band 100–2500 Hz (see Equation (4)).

$$\tau_{\text{avg}} = \frac{1}{n}\sum_{i=1}^{n}\tau_f, \tag{4}$$

where n is the number of frequency bands.

To calculate the average transmission loss in decibels, the following Equation (5) is used:

$$TL_{\text{avg}} = -10\log_{10}\left(\tau_{\text{avg}}\right). \tag{5}$$

This formula allows converting the average transmission coefficient τ_f into the corresponding transmission loss in decibels.

When comparing samples, A and B, the transmission coefficient of power of both samples will be evaluated using the relative transmission ratio (RTR_f), for which the following Equation (6) applies:

$$RTR_f = \frac{\tau_{f,vzorkaA}}{\tau_{f,vzorkaB}}, \tag{6}$$

where $\tau_{f,samplesA}$ is the power transmission coefficient of sample A at frequency f a $\tau_{f,samplesB}$ is the power transmission coefficient of sample A at the same frequency. The power transmission ratio indicates how many times the power transmission coefficient of one sample is greater or smaller than that of another. Interpretation of the RTR_f value:

1. If $RTR_f > 1$, sample A transmits more input power at the given frequency than sample B. This indicates sample A has lower transmission loss (less power loss) compared to sample B.
2. If $RTR_f < 1$, sample A transmits less input power at the given frequency than sample B. This means sample A has greater transmission loss (more power loss) compared to sample B.
3. If $RTR_f \approx 1$, both samples transmit nearly the same amount of input power at the given frequency, indicating similar transmission loss between the two.

2.5.2. Regression Analysis

To determine the relationship between variables, we consider variable Y, which represents the dependent (output) variable, and the variables $X_1,X_2,\ldots,X_k$, which represent k-independent (input) variables. The relationship between the dependent variable Y and independent variables X_i (where $i = 1,\ldots,k$) can be described using the following multiple linear regression model (see Equation (7)):

$$Y = \beta_0 + \sum_{i=1}^{k} \beta_i X_i + \varepsilon, \tag{7}$$

where β_0 is the intercept, $\beta_1,\beta_2,\ldots,\beta_k$ are the regression coefficients corresponding to each independent variable, and ε is the error term, representing the variability not explained by the model.

The statistical significance of the entire regression model can be tested using the F-test. To verify the statistical significance of individual parameters of the regression model, tests of statistical significance of regression coefficients, most often t-tests, are used. When testing statistical hypotheses, the decision to reject or accept the null hypothesis is implemented based on the p-value. If the p-value is less than the established significance level α, we reject the null hypothesis in favor of the alternative hypothesis. If the p-value value is less than the established significance level α, we reject the null hypothesis in favor of the alternative hypothesis. If the p-value is equal to or greater than the chosen significance level of α, then we do not reject the null hypothesis.

The strength of the dependence of the variable Y on the effect on the independent variables is expressed by a multiple coefficient of determination R^2, which has values in the interval $\langle 0; 1 \rangle$. The closer the value of R^2 is to 1, the closer the relationship between the dependent and the independent variables, and the model explains more of the dependent variable's variability. Because variables in regression analysis can be expressed in different units and scales, standardized regression coefficients β^*_i, where $i = 1,\ldots,k$, are often used to compare the relative importance of individual input variables. The standardized regression coefficient β^*_i expresses how many standard deviations the output variable Y changes when the standardized deviation of the independent variable X_i changes by one unit, assuming that the other input variables remain constant. The higher the absolute value of the coefficient β^*_i, the stronger the influence of the independent variable in question on the output variable.

3. Results

In this section, we present a comprehensive analysis of the acoustic properties of recycled rubber crumb, with a particular emphasis on sound transmission loss. The study investigates how various sample parameters affect the material's ability to dampen sound, focusing on four main variables: rubber crumb fraction (ranging from fine particles of 1 mm to coarse particles of 3–4 mm); sample thickness (2 cm, 4.5 cm, and 7 cm); compression pressure of the material (0 kPa—loose granular material, 250, 500, and 750 kPa—granular material); sound frequency (ranging from 100 Hz to 2500 Hz).

Methodologically, we compare the sound transmission loss and transmission coefficient of compacted samples with those of loose granular rubber crumb under various configurations of these parameters. For each sample, under different configurations of the specified parameters (pressure, thickness, and fraction), five repeated measurements of transmission loss were performed. To enable comparison of the observed properties, the average transmission loss was calculated at each frequency within the considered frequency range. The results provide a systematic evaluation of the measured values, aiming to identify the optimal properties of recycled rubber crumb for maximizing its acoustic insulation capabilities. The insights gained from this research have the potential to support more efficient use of recycled rubber materials in acoustic applications, thereby promoting environmentally friendly waste management while enhancing sound insulation in buildings, industrial plants, and other environments where noise reduction is essential.

3.1. Defining Variables

Recycled tire-derived rubber crumb demonstrates distinctive acoustic properties that make it suitable for sound insulation and absorption applications. Studies indicate its effectiveness in noise reduction, attributed to its porous internal structure [34–37].

Table 2 provides a systematic overview of the material's properties, experimental parameters, and variables characterizing the analyzed rubber crumb samples.

Table 2. Overview of the parameters used and properties of the examined samples.

Label	Description
F	Rubber material fraction (mm)
F1	Fraction of rubber material of size 1 mm (Fraction 1)
F2	Fraction of rubber material of size from 1–2 mm (Fraction 2)
F3	Fraction of rubber material of size from 2–3 mm (Fraction 3)
F4	Fraction of rubber material of sizes from 3–4 mm (Fraction 4)
h	Sample thickness (cm), h = 2 cm, h = 4.5 cm, h = 7 cm
p	Pressure under which the granular sample was compacted (kPa), 0 kPa (loose granular), 250 kPa, 500 kPa, 750 kPa (granular material)
Freq	Frequency (Hz), frequency band from 100 Hz to 2500 Hz
TL	Transmission loss (dB)
TL_C	Transmission loss of compact material of recycled rubber crumb
TL_F	Transmission loss of granular recycled rubber crumb at fraction *F*
$TL_{C,h}$	Transmission loss of compact material of recycled rubber crumb at thickness *h*
$TL_{F,p,h}$	Transmission loss of granular recycled rubber crumb at fraction *F*, pressure *p* and thickness *h*
τ_f	Transmission coefficient at frequency *f* (unitless, –)
τ_{avg}	Average transmission coefficient (unitless, –)
TL_f	Transmission loss at frequency *f* (dB)
TL_{avg}	Average transmission loss (dB)
RTR_f	Relative transmission ratio at frequency *f* (unitless, –)

3.2. Acoustic Performance Evaluation of Recycled Rubber Crumb

3.2.1. Transmission Loss and Transmission Coefficient Analysis

The compact recycled rubber crumb (RRC) was produced in three different thicknesses: 2 cm, 4.5 cm, and 7 cm. The best transmission loss values (in dB) were achieved by the sample with a thickness of 7 cm ($TL_{C,7}$). The highest values were measured in the 7 cm sample within the frequency range of 1600 Hz to 2500 Hz, with the maximum transmission loss, $TL_{C,7}$ = 33.13 dB, recorded at a frequency of 2000 Hz (see Figure 4).

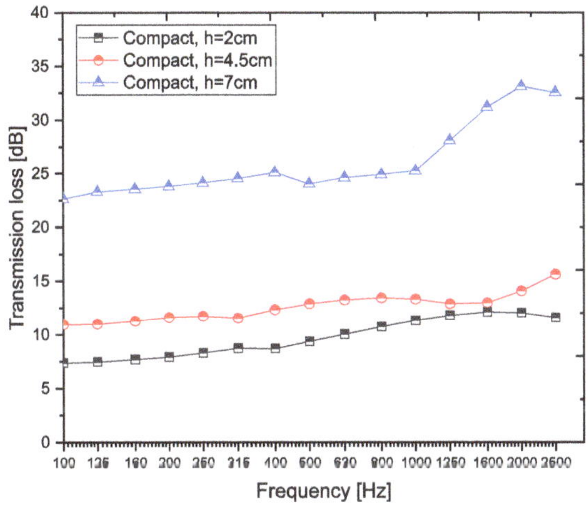

Figure 4. Transmission loss (TL) values compact recycled rubber granulate (RRC) samples at different thicknesses across frequencies.

A graphical representation of the transmission loss values for samples of granular recycled rubber crumb, including the effects of particle size fraction, compaction pressure, and thickness, is presented in Figure 5. To enhance the informative value, the graphs also include comparative data for compact recycled rubber crumb at equivalent sample thicknesses, enabling direct analysis of the differences between granular and compact forms of the material.

For the 2 cm thick sample (Figure 5), the highest transmission loss (TL, in dB) values were observed in *Fraction 1* under compaction pressures of 500 kPa and 750 kPa. Conversely, the lowest TL values occurred in *Fraction 4*, with minimal variation across pressure levels. Notably, *Fractions 3* and *4* (particle sizes 2–4 mm) exhibited nearly identical TL values, likely due to their small particle size differences and identical material composition.

Compared to the transmission loss values for *Compact RRC* and *Fraction 1*, the following applies: $TL_{F1,2,0} < TL_{C,2} < TL_{F1,2,250} < TL_{F1,2,500} < TL_{F1,2,750}$. For *Fraction 2* it applies that $TL_{F2,2,0} < TL_{F2,2,250} < TL_{C,2} < TL_{F2,2,500} < TL_{F2,2,750}$, and for *Fraction 3*, or *Fraction 4* it applies that $TL_{F3,2,0} < TL_{F3,2,250} < TL_{F3,2,500} < TL_{F3,2,750} < TL_{C,2}$, resp. $TL_{F4,2,0} < TL_{F4,2,250} < TL_{F4,2,500} < TL_{F4,2,750} < TL_{C,2}$.

Table 3 shows the transmission coefficient τ_f at frequency f, expressed as a percentage, for samples with a thickness of 2 cm at various particle size fractions and compaction pressures across the entire frequency range, including data for *Compact RRC* of the same thickness. To compare the efficiency of the individual samples, the table also presents the average linear transmission coefficient τ_{avg} (in %) and the average transmission loss TL_{avg} (in dB), both calculated over the full frequency spectrum.

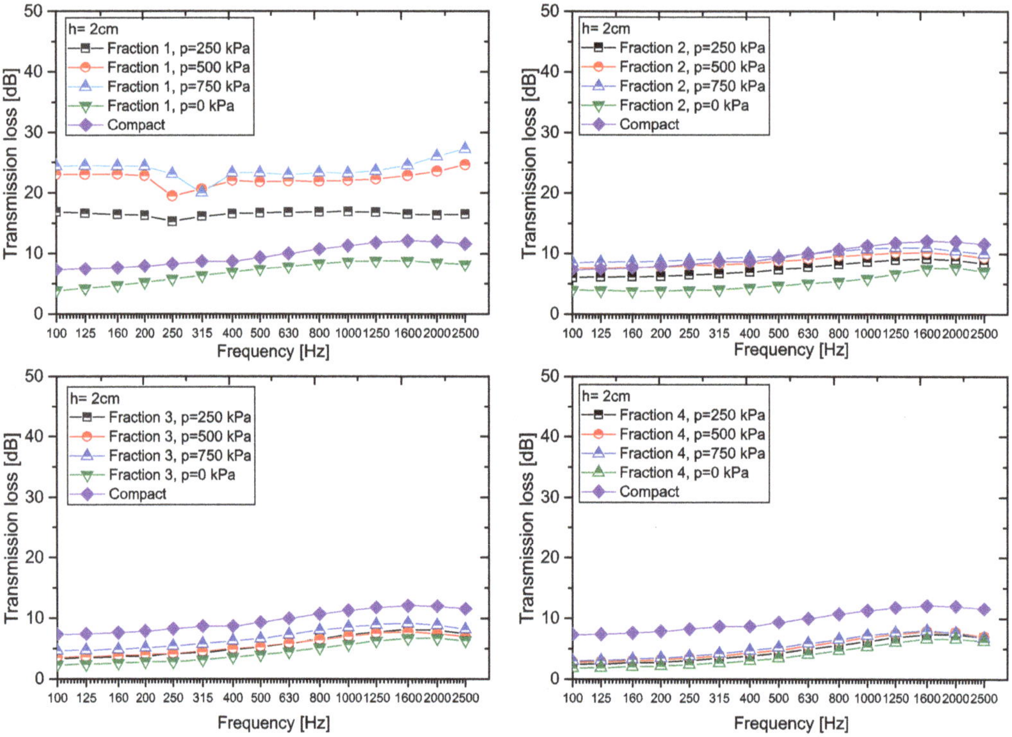

Figure 5. Graphical representation of the measured TL values for RRC samples (h = 2 cm).

The analysis reveals that for *Fraction 1* under a pressure of 250 kPa at a frequency of 100 Hz, only approximately 2.1% of the original sound power remains, meaning the transmission coefficient is 2.1%. In contrast, for *Fraction 4* at a pressure of 0 kPa and the same frequency, as much as 64.8% of the original power remains, indicating significantly lower transmission loss compared to *Fraction 1*. *Fraction 1* samples achieve the lowest transmission coefficients at pressures of 500 kPa and 750 kPa, while *Fraction 4* samples consistently show the highest transmission coefficients. For each fraction, samples made from loose granular recycled rubber crumb (p = 0 kPa) exhibit the highest transmission coefficients. In most cases, except for *Fraction 1* at pressures of 250 kPa and 500 kPa, the highest transmission coefficient is recorded at the lowest frequency (100 Hz), with values generally decreasing as frequency increases. The most pronounced differences between the lowest and highest frequencies are observed in *Fraction 3* and *Fraction 4*. For the *Fraction 2*, *Fraction 3*, and *Fraction 4* samples, the transmission coefficient decreases up to a frequency of 1000 Hz, after which it increases, resulting in reduced attenuation effects at higher frequencies.

The experimental results indicate that, to achieve maximum sound insulation, Fraction 1 should be used under pressure (at or above 500 kPa), as this configuration provides the greatest attenuation effect across a wide frequency spectrum. For specialized applications that require the suppression of sound waves within a specific frequency range, combinations involving larger fractions with variable pressure profiles can also be effective alternatives.

For a 2 cm thick *Compact RRC* sample, the average transmission coefficient τ_{avg} (in%) for the 100–2500 Hz frequency band was determined to be 11.62% (see Table 3). In contrast, graphical analysis (see Figure 6) shows the average transmission coefficient in the same frequency range for granular *RRC* with a thickness of 2 cm, across all investigated fractions.

Table 3. Transmission coefficient for samples with a thickness of 2 cm at frequency *f* (in %).

Freq (Hz)	F1				F2				Compact
	250 kPa	500 kPa	750 kPa	0 kPa	250 kPa	500 kPa	750 kPa	0 kPa	
100	2.1	0.5	0.4	40.8	24.9	17.1	13.7	39.7	18.3
125	2.2	0.5	0.4	37.8	24.8	17.0	13.6	40.7	17.9
160	2.3	0.5	0.4	33.4	24.2	16.7	13.4	41.5	17.1
200	2.4	0.5	0.4	29.8	23.6	16.3	13.1	41.2	16.1
250	3.0	1.1	0.5	26.4	22.8	15.6	12.6	40.5	14.8
315	2.4	0.9	1.0	23.1	21.7	14.9	12.0	39.1	13.4
400	2.2	0.6	0.5	20.3	20.2	14.2	11.2	36.8	13.5
500	2.1	0.7	0.5	18.1	18.7	13.3	10.9	33.8	11.6
630	2.1	0.6	0.5	16.2	16.9	12.1	9.8	30.9	9.9
800	2.1	0.6	0.5	14.5	15.0	11.0	8.9	28.9	8.4
1000	2.0	0.6	0.5	13.5	13.6	10.2	8.3	25.7	7.4
1250	2.1	0.6	0.4	13.1	12.6	9.6	7.9	21.5	6.6
1600	2.2	0.5	0.3	13.2	12.1	9.4	7.9	17.7	6.1
2000	2.3	0.4	0.2	14.0	12.8	10.1	8.8	17.0	6.3
2500	2.2	0.3	0.2	15.0	14.4	11.5	10.0	19.6	7.0
τ_{avg} (in %)	2.25	0.60	0.43	21.96	18.57	13.27	10.81	31.64	11.62
TL_{avg} (in dB)	16.48	22.22	23.67	6.58	7.31	8.77	9.66	5.00	9.35
Freq (Hz)	**F3**				**F3**				**Compact**
	250 kPa	**500 kPa**	**750 kPa**	**0 kPa**	**250 kPa**	**500 kPa**	**750 kPa**	**0 kPa**	
100	45.9	43.3	33.7	58.2	57.5	52.0	49.2	64.8	18.3
125	45.0	42.6	33.1	56.9	56.6	51.1	48.3	64.3	17.9
160	43.4	41.0	31.8	54.3	54.7	49.3	46.3	62.3	17.1
200	41.5	39.2	30.2	52.0	52.4	47.0	43.9	60.1	16.1
250	38.6	37.7	28.1	51.5	49.1	44.0	40.9	57.2	14.8
315	36.4	34.8	25.7	47.4	45.3	40.5	37.2	53.8	13.4
400	33.3	31.8	22.9	43.0	41.5	36.3	33.3	49.2	13.5
500	29.4	28.9	21.1	39.7	37.0	32.6	29.6	44.4	11.6
630	25.8	25.6	18.2	35.4	32.3	28.3	25.6	39.3	9.9
800	21.9	22.2	15.5	30.4	27.5	24.0	21.8	33.6	8.4
1000	19.0	19.4	13.7	26.5	23.6	20.6	19.0	28.8	7.4
1250	16.7	17.4	12.5	23.4	20.5	17.9	16.9	24.8	6.6
1600	15.3	16.6	11.9	21.1	18.2	16.2	15.8	21.7	6.1
2000	15.9	17.6	12.8	21.1	18.2	16.9	17.1	21.5	6.3
2500	17.9	19.7	14.9	23.1	20.4	19.8	20.7	23.8	7.0
τ_{avg} (in %)	29.74	29.20	21.74	38.95	36.98	33.11	31.03	43.31	11.62
TL_{avg} (in dB)	5.27	5.35	6.63	4.09	4.32	4.80	5.08	3.63	9.35

In the case of sample *Fraction 1* (thickness 2 cm, pressure 250 kPa), the average transmission coefficient τ_{avg} = 2.2%, which means that the material allows only 2.2% of the incident energy to pass through, and the rest (97.7%) is absorbed or lost due to absorption or reflection (Figure 6). In comparison, for the same fraction with zero pressure, τ_{avg} increases to 22%, demonstrating a significant effect of the compaction of the sample under pressure on the attenuation properties of the material. The highest τ_{avg} values are shown by *Fraction 4* (31.0–43.3%) and *Fraction 3* (21.2–39.3%) samples, with ranges reflecting the dependence on the applied pressure. These results confirm that higher *Fractions (3–4)* show lower attenuation in the given frequency band.

For samples with a thickness of 4.5 cm, *Fraction 1* samples achieved the highest transmission loss values at pressures of 500 kPa and 750 kPa, while at approximately 400 Hz, the transmission loss values began to increase (see Figure 7). Also in this case, the transmission loss values for *Fraction 3* and *Fraction 4* samples are very similar, which indicates comparable acoustic properties of these fractions regardless of the increased sample thickness.

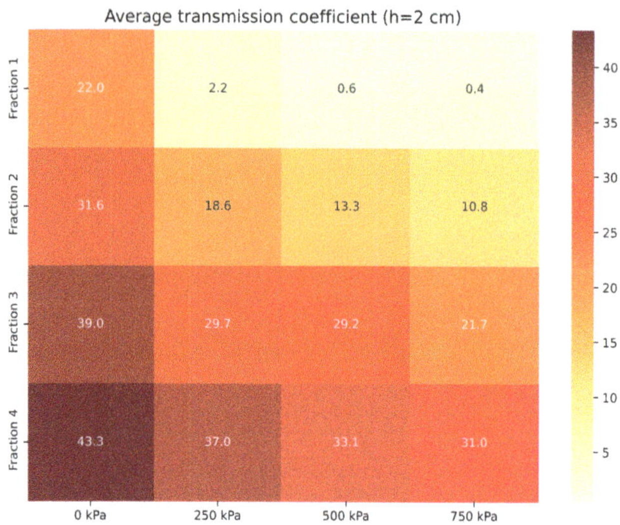

Figure 6. Average transmission coefficient τ_{avg} of granular RRC (h = 2 cm).

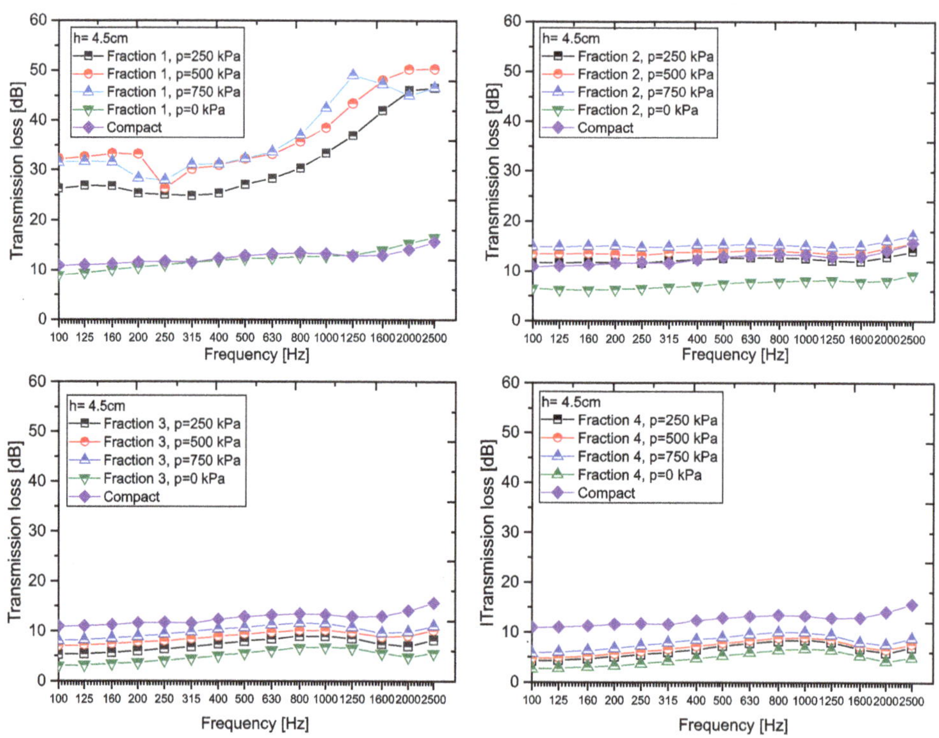

Figure 7. Graphical representation of the measured TL values for RRC samples (h = 4.5 cm).

When comparing the transmission loss values for all samples with a thickness of 4.5 cm, the following trends can be observed. For *Fraction 1*, the relationship is as follows: $TL_{F1,4.5,0} < TL_{C,4.5} < TL_{F1,4.5,250} < TL_{F1,4.5,500} < TL_{F1,4.5,750}$. For *Fraction 2* it applies that $TL_{F2,4.5,0} < TL_{C,4.5} < TL_{F2,4.5,250} < TL_{F2\ 4.5,500} < TL_{F2,4.5,750}$, and for *Fraction 3*, or *Fraction 4* it applies that $TL_{F3,4.5,0} < TL_{F3,4.5,250} < TL_{F3,4.5,500} < TL_{F3,4.5,750} < TL_{C,4.5}$, or $TL_{F4,4.5,0} < TL_{F4,4.5,250} < TL_{F4,4.5,500} < TL_{F4,4.5,750} < TL_{C,4.5}$.

Table 4 shows the transmission coefficient τ_f (in %) for samples with a thickness of 4.5 cm at different fractions and pressures throughout the frequency band, including the *Compact RRC* of the same thickness. To compare the efficiency of individual samples, the table also shows the average linear transmission coefficient τ_{avg} (in %) and the average transmission loss TL_{avg} (in dB), calculated for the entire frequency spectrum.

Table 4. Transmission coefficient for samples with a thickness of 4.5 cm at frequency *f* (in %).

Freq (Hz)	F1				F2				Compact
	250 kPa	500 kPa	750 kPa	0 kPa	250 kPa	500 kPa	750 kPa	0 kPa	
100	0.23	0.06	0.07	12.93	6.83	4.53	3.20	22.63	7.98
125	0.21	0.05	0.07	11.46	6.77	4.54	3.19	23.65	7.91
160	0.21	0.05	0.07	9.93	6.63	4.45	3.14	24.07	7.41
200	0.29	0.05	0.14	8.79	6.65	4.62	3.08	23.53	6.87
250	0.31	0.23	0.16	7.82	6.90	4.74	3.32	22.67	6.70
315	0.33	0.10	0.08	7.03	6.18	4.23	3.23	21.35	6.97
400	0.29	0.08	0.08	6.39	5.80	4.07	2.96	19.60	5.82
500	0.19	0.06	0.06	6.01	5.56	3.94	2.92	17.88	5.09
630	0.15	0.05	0.04	5.70	5.34	3.84	2.86	16.80	4.72
800	0.09	0.03	0.02	5.46	5.29	3.85	2.90	16.48	4.52
1000	0.05	0.01	0.01	5.27	5.57	4.07	3.07	15.57	4.66
1250	0.02	0.005	0.001	4.90	6.149	4.41	3.29	15.22	5.14
1600	0.01	0.002	0.002	3.97	6.339	4.32	3.11	16.35	5.02
2000	0.002	0.001	0.003	2.89	5.093	3.39	2.39	15.53	3.90
2500	0.002	0.001	0.002	2.21	3.941	2.73	1.91	11.72	2.73
τ_{avg} (in %)	0.159	0.051	0.053	6.72	5.94	4.12	2.97	18.87	5.70
TL_{avg} (in dB)	28.00	32.89	32.77	11.73	12.27	13.86	15.27	7.24	12.44
Freq (Hz)	F3				F4				Compact
	250 kPa	500 kPa	750 kPa	0 kPa	250 kPa	500 kPa	750 kPa	0 kPa	
100	29.51	19.57	15.40	48.96	37.91	33.39	26.04	54.23	7.98
125	28.68	18.97	14.76	47.73	36.48	31.93	24.68	52.50	7.91
160	26.89	17.87	13.78	45.10	33.82	29.48	22.75	49.59	7.41
200	25.18	16.72	12.74	42.29	31.07	27.01	20.75	46.20	6.87
250	23.04	15.78	11.49	38.69	27.91	24.29	18.35	42.42	6.70
315	20.71	14.35	10.24	34.90	25.17	21.72	15.89	38.02	6.97
400	18.12	12.80	9.05	30.86	21.84	18.76	14.06	33.31	5.82
500	15.98	11.55	8.52	27.44	18.80	16.30	12.54	29.32	5.09
630	14.24	10.49	7.45	24.20	16.47	14.25	10.81	25.44	4.72
800	12.86	9.72	6.95	21.71	14.78	12.84	9.81	22.41	4.52
1000	12.72	9.72	7.09	20.66	14.32	12.58	9.82	21.12	4.66
1250	14.164	10.71	8.30	21.88	15.72	13.90	11.33	22.32	5.14
1600	18.323	12.96	10.78	27.54	21.25	18.84	15.75	29.17	5.02
2000	19.898	12.35	10.31	32.89	25.49	21.97	17.25	37.62	3.90
2500	15.241	9.65	7.89	27.57	20.05	17.01	13.01	32.28	2.73
τ_{avg} (in %)	19.70	13.35	10.32	32.83	24.07	20.98	16.19	35.73	5.70
TL_{avg} (in dB)	7.05	8.68	9.86	4.84	6.18	6.79	7.91	4.47	12.44

Also in this case, *Fraction 1* samples show the lowest transmission coefficient at pressures of 500 kPa and 750 kPa, which means that these samples achieve the highest transmission loss. On the contrary, the highest transmission coefficient (i.e., the lowest transmission loss) was found in *Fraction 4* samples. For all fractions, samples made from loose granular (p = 0 kPa) recycled rubber crumb exhibit the highest transmission coefficient, indicating a lower ability to dampen acoustic energy compared to compacted samples.

In all cases, the highest transmission coefficient occurs at the lowest frequency, i.e., at 100 Hz. For *Fraction 2*, *Fraction 3*, and *Fraction 4* samples, the transmission coefficient decreases with increasing frequency up to 1000 Hz. Above this frequency, there is a resurgence of the transmission coefficient, which means a decrease in attenuation effects in this part of the frequency spectrum. These results confirm the importance of the combination of fraction size, applied pressure, and material thickness for optimizing the acoustic properties of the tested samples.

The average transmission coefficient τ_{avg} (in %) was determined to be 5.70 % in this case for a *Compact RRC* sample with a thickness of 4.5 cm (see Table 4). Figure 8 shows a graphical analysis of τ_{avg} (in %) for a granular RRC sample of the same thickness (h = 4.5 cm).

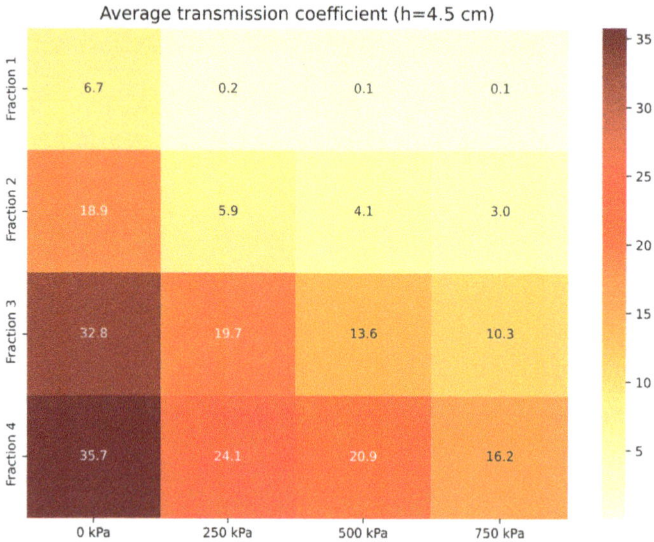

Figure 8. Average transmission coefficient τ_{avg} of granular RRC (h = 4.5 cm).

For *Fraction 1* (4.5 cm thickness, excluding loose granular RRC, p = 0 kPa), the average transmission coefficient (τ_{avg}) was measured at 0.1%, reflecting exceptional sound attenuation. The highest values of the average transmission coefficient were again recorded for the samples *Fraction 3* and *Fraction 4*—loose granular RRC pressure 0 kPa.

Experimental results confirm that the *Fraction 1* sample at the applied pressure of 750 kPa shows the highest average transmission loss TL_{avg}, which ranks it among the materials with the most effective acoustic insulation properties (see Figure 9). In comparison, *Fraction 4* at pressure 0 kPa achieves the lowest damping efficiency, which means a low ability to suppress the propagation of sound energy.

When comparing the transmission loss values for all samples with a thickness of 7 cm, the following trends can be observed. For *Fraction 1*, the relationship is as follows: $TL_{F1,7,0} < TL_{C,7} < TL_{F1,7,250} < TL_{F1,7,500} < TL_{F1,7,750}$. For *Fraction 2* it applies that $TL_{F2,7,0} < TL_{F2,7,250} < TL_{C,7} < TL_{F2,7,500} < TL_{F2,7,750}$, and for samples *Fraction 3*, or *Fraction 4* it applies that $TL_{F3,7,0} < TL_{F3,7,250} < TL_{F3,7,500} < TL_{F3,7,750} < TL_{C,7}$, or $TL_{F4,7,0} < TL_{F4,7,250} < TL_{F4,7,500} < TL_{F4,7,750} < TL_{C,7}$.

Table 5 presents the transmission coefficient τ_f (in %) for samples with a thickness of 7 cm at different fractions and pressures throughout the frequency band, including

Compact RRC of the same thickness. In addition, the table also includes the average linear transmission coefficient τ_{avg} (in %) and the average transmission loss TL_{avg} (in dB). All these values provide an overview of the acoustic properties of individual samples.

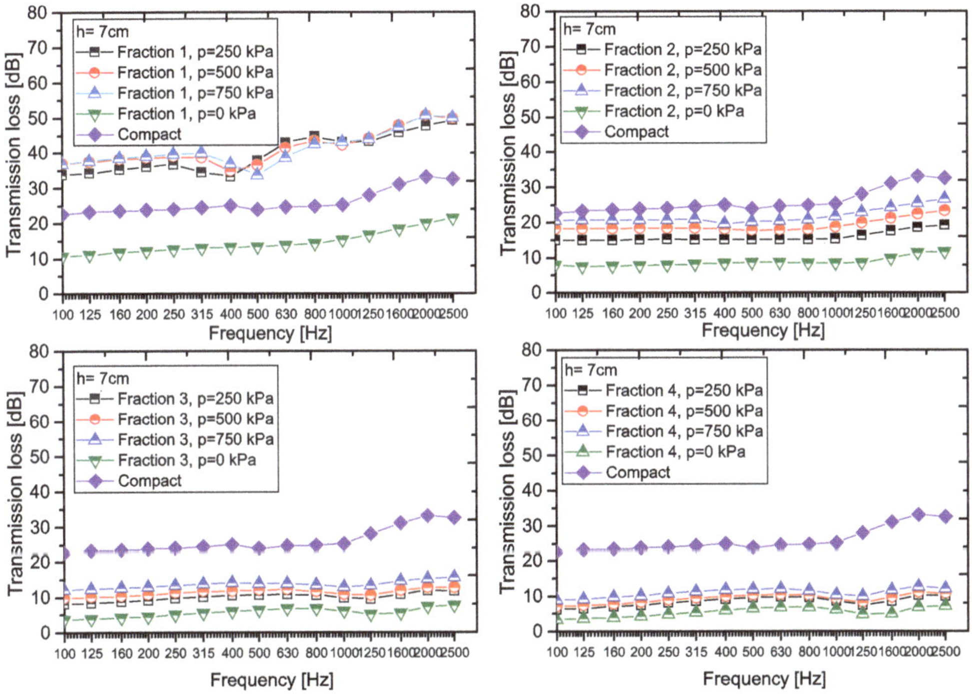

Figure 9. Graphical representation of the measured TL values for RRC samples (h = 7 cm).

Fraction 1 samples at pressures (500–750 kPa) show an almost zero average transmission coefficient ($\tau_{avg} \approx 0\%$), which ranks them among materials with excellent acoustic properties. These results confirm their effectiveness in suppressing the transmission of acoustic energy throughout the researched frequency band. Conversely, *Fraction 4* at p = 0 kPa achieves the highest values of τ_{avg}, which indicates its reduced ability to suppress the transmission of acoustic energy. *Fraction 2* and *Compact RRC* show significantly higher attenuation at a thickness of 7 cm compared to previous experimental test conditions, which makes them effective acoustic materials. The thickness of the material is of great importance in this case. When the thickness of the sample is increased to 7 cm, the cut-off frequency is shifted to the region of low frequencies (100–500 Hz). In this band, the material begins to resonate more strongly (waves "accumulate" and amplify in it). This behavior causes the sound waves in the material to slow down and bounce off the edges, creating standing waves.

Experimental data have shown that the 7 cm thick *Compact RRC* has an average transmission coefficient τ_{avg} = 0.31% (see Table 5), confirming its almost complete impermeability to sound waves. For the loose granular (p = 0 kPa) sample with a thickness of 7 cm (Figure 10), τ_{avg} was the lowest compared to thicknesses of 2 and 4.5 cm, indicating the synergistic effect of the combination of compaction and sample thickness. At the same time, *Fraction 1* (except at p = 0 kPa) almost completely prevents the transmission of incident power.

These results show that optimizing sound attenuation requires a multifactorial approach—a combination of optimal particle size, sufficient material thickness, and precise pressure adjustment. The synergistic effect of these parameters determines how effectively the material suppresses the propagation of sound energy, while the interaction between the material structure and the input parameters is key to achieving the desired acoustic properties.

Table 5. Transmission coefficient for samples with a thickness of 7 cm at frequency f (in %).

Freq (Hz)	F1				F2				Compact
	250 kPa	500 kPa	750 kPa	0 kPa	250 kPa	500 kPa	750 kPa	0 kPa	
100	0.04	0.02	0.02	8.6	3.2	1.5	0.9	16.3	0.5
125	0.04	0.02	0.02	7.6	3.2	1.5	0.8	17.4	0.5
160	0.03	0.02	0.01	6.7	3.1	1.5	0.8	17.2	0.4
200	0.02	0.01	0.01	6.0	3.0	1.4	0.8	16.8	0.4
250	0.02	0.01	0.01	5.5	2.9	1.4	0.8	16.1	0.4
315	0.04	0.01	0.01	5.1	3.1	1.4	0.8	15.2	0.4
400	0.05	0.03	0.02	4.8	3.1	1.5	1.1	14.1	0.3
500	0.02	0.02	0.04	4.5	3.0	1.7	0.9	13.3	0.4
630	0.005	0.01	0.01	4.2	3.1	1.6	0.9	13.2	0.3
800	0.003	0.00	0.01	3.7	3.1	1.6	0.8	14.0	0.3
1000	0.005	0.01	0.005	2.9	2.9	1.3	0.7	14.2	0.3
1250	0.005	0.004	0.004	2.2	2.3	1.0	0.5	13.8	0.2
1600	0.003	0.002	0.002	1.4	1.7	0.8	0.4	10.3	0.1
2000	0.002	0.001	0.001	1.0	1.4	0.6	0.3	7.2	0.0
2500	0.001	0.001	0.001	0.7	1.2	0.5	0.2	6.8	0.1
τ_{avg} (in %)	0.018	0.011	0.012	4.33	2.69	1.28	0.71	13.74	0.31
TL_{avg} (in dB)	37.34	39.42	39.29	13.64	15.69	18.92	21.49	8.62	25.14
Freq (Hz)	F3				F4				Compact
	250 kPa	500 kPa	750 kPa	0 kPa	250 kPa	500 kPa	750 kPa	0 kPa	
100	15.5	10.4	6.1	43.1	23.8	19.4	12.6	45.2	0.5
125	14.7	9.9	5.8	41.3	22.3	18.3	11.7	43.2	0.5
160	13.5	9.2	5.3	38.1	20.1	16.4	10.5	39.5	0.4
200	12.3	8.4	4.9	35.1	17.9	14.7	9.4	36.5	0.4
250	10.9	7.5	4.5	31.6	15.6	12.7	8.2	32.4	0.4
315	9.9	6.9	4.1	28.2	13.6	11.2	7.2	28.3	0.4
400	9.1	6.6	3.8	25.0	12.3	10.2	6.3	24.7	0.3
500	8.6	6.1	4.0	22.9	11.0	9.2	6.2	22.1	0.4
630	8.3	6.1	4.0	21.2	10.4	8.8	6.1	20.2	0.3
800	8.9	6.7	4.4	21.3	11.2	9.6	6.9	20.3	0.3
1000	10.5	8.2	5.0	24.3	13.9	12.1	8.8	23.6	0.3
1250	11.5	8.8	4.6	30.2	18.1	15.0	9.9	31.9	0.2
1600	8.5	6.4	3.4	28.2	14.1	10.9	6.7	31.4	0.1
2000	6.6	5.3	2.9	18.8	9.6	7.7	5.4	20.2	0.0
2500	6.7	5.4	2.7	17.0	10.0	8.5	6.1	18.7	0.1
τ_{avg} (in %)	10.38	7.46	4.39	28.41	14.93	12.31	8.14	29.2	0.31
TL_{avg} (in dB)	9.84	11.27	13.58	5.46	8.26	9.10	10.89	5.335	25.14

3.2.2. Comparison of Samples Using the Relative Transmission Ratio (RTR)

For the direct comparison of the acoustic properties of compact and loose granular RRC samples at a given frequency f, the relative transmission ratio (RTR_f) is determined by the following equation (Equation (5)):

$$RTR_f = \frac{\tau_{f,\ compact}}{\tau_{f,\ granular}}, \tag{8}$$

where $\tau_{f,compact}$ represents the transmission coefficient of a *Compact RRC* sample at frequency f, and $\tau_{f,granular}$ denotes the transmission coefficient of granular RRG (at pressures ranging from 0 to 750 kPa) at the same frequency f, with both samples having the same thickness.

A graphical representation of the RTR_f values (see Equation (8)) for each sample thickness and each frequency f is shown in Figures 11–14.

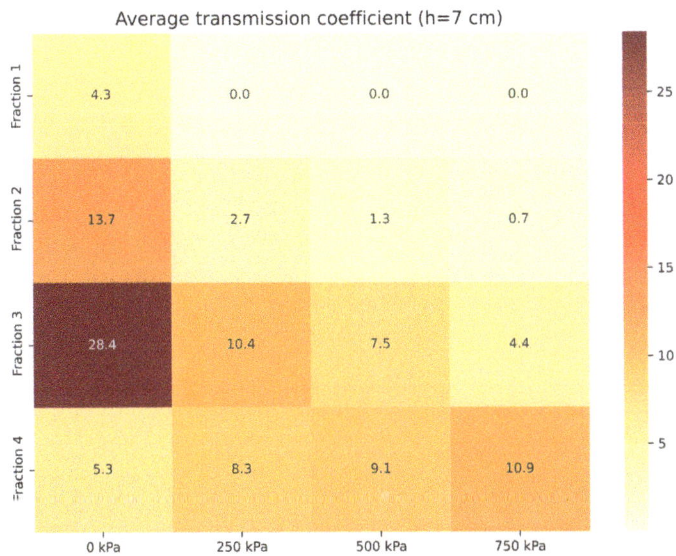

Figure 10. Average transmission coefficient τ_{avg} of granular RRC (h = 7 cm).

The greatest variability of RTR values was recorded when comparing compact and granular recycled rubber crumb (p = 0–750 kPa) of *Fraction 1* (h = 2 cm). In all cases, except for the loose granular sample (p = 0 kPa), the RTR values were several times higher than 1, which means that the *Compact RRC* transmits several times more incident power than the granular crushed material. In the case of *Fraction 2*, RTR values are greater than 1 at a pressure of 750 kPa in the frequency range from 100 Hz to 800 Hz, with values approaching 1 above this limit. At a pressure of 500 kPa and in the band of 100 to 315 Hz, the transmission coefficient τ_f and thereby the transmission loss of TL_f of compact and granular samples of *Fraction 2* are comparable.

For *Fraction 3* and *Fraction 4* samples, the RTR values are consistently below 1, indicating that the *Compact RRC* transmits less incident power than the granular RRC (p = 0–750 kPa) at the same frequency and thickness. This corresponds to a significantly higher transmission loss for the *Compact RRC* compared to the granular RRC, with the transmitted power through the *Compact RRC* sample being, on average, 2.5 to 3.9 times lower than that of the granular RRC.

A graphical representation of the relative transmission ratio value for samples 4.5 cm thick is shown in Figures 12 and 13.

The greatest variability in the relative transmission ratio (RTR) values again occurs when comparing compact and granular recycled rubber crumb (p = 0–750 kPa) of *Fraction 1* (h = 4.5 cm). In all cases, except for loose granular crumb (p = 0 kPa), the RTR values are several times higher than 1. For better clarity, Figure 13 shows ratios from 100 Hz and above. This indicates that granular RRC has a significantly better transmission loss at pressures (p = 250–750 kPa) compared to *Compact RRC*. In the case of *Fraction 2*, RTR values are greater than 1 at pressures of 500 kPa and 750 kPa, indicating a higher efficiency of granular RRC in this range. At a pressure of 250 kPa, the relative transmission ratio value is close to 1, which means that the transfer properties of both types of samples are comparable. For *Fraction 3* and *Fraction 4* samples, the RTR values are balanced and less than 1, which means that the sample from the *Compact RRC* passed through less incident power than the sample of the granular RRC at the same frequency and thickness. This corresponds to higher attenuation for the *Compact RRC* in these cases.

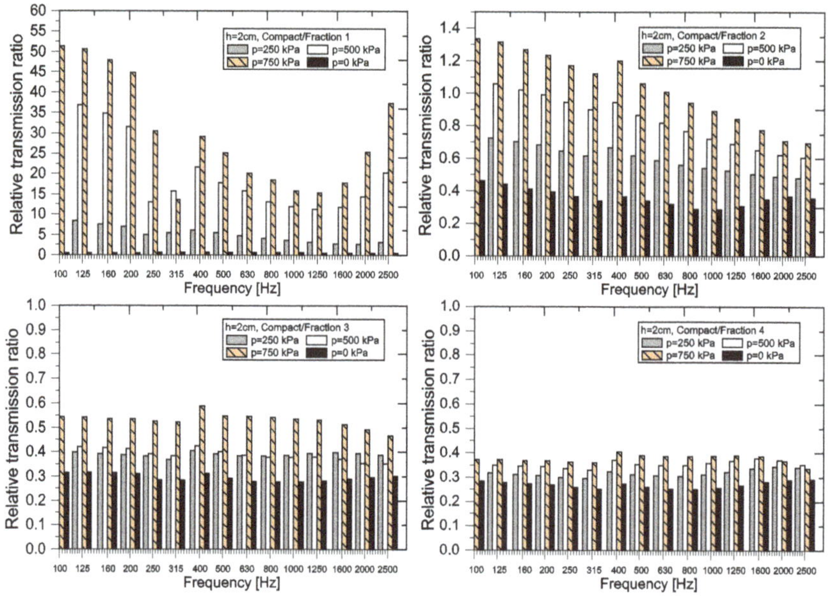

Figure 11. Graphical representation of RTR for RRC samples (h = 2 cm).

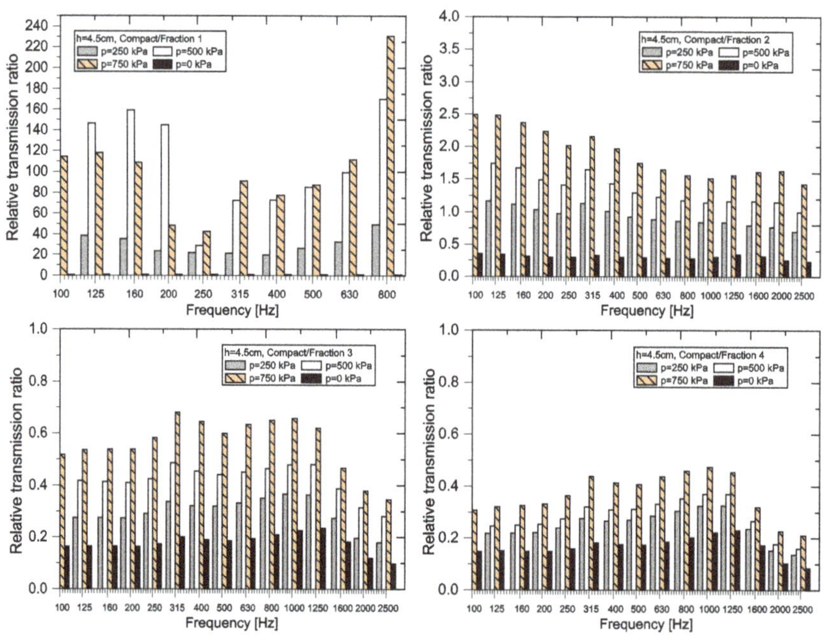

Figure 12. Graphical representation of RTR for RRC samples (h = 4.5 cm).

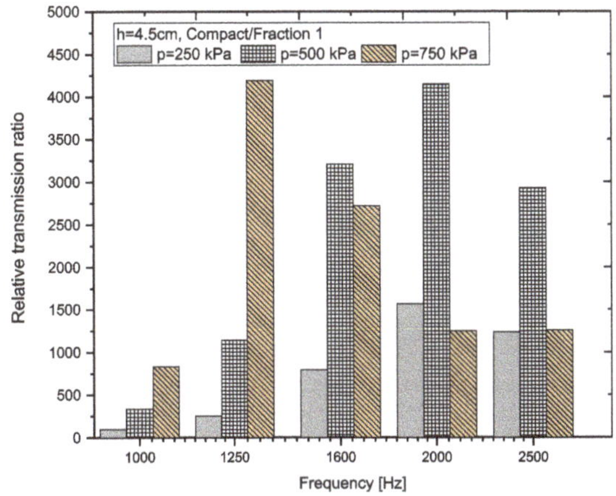

Figure 13. Graphical representation of RTR for RRC samples (Fraction 1, h = 4.5 cm).

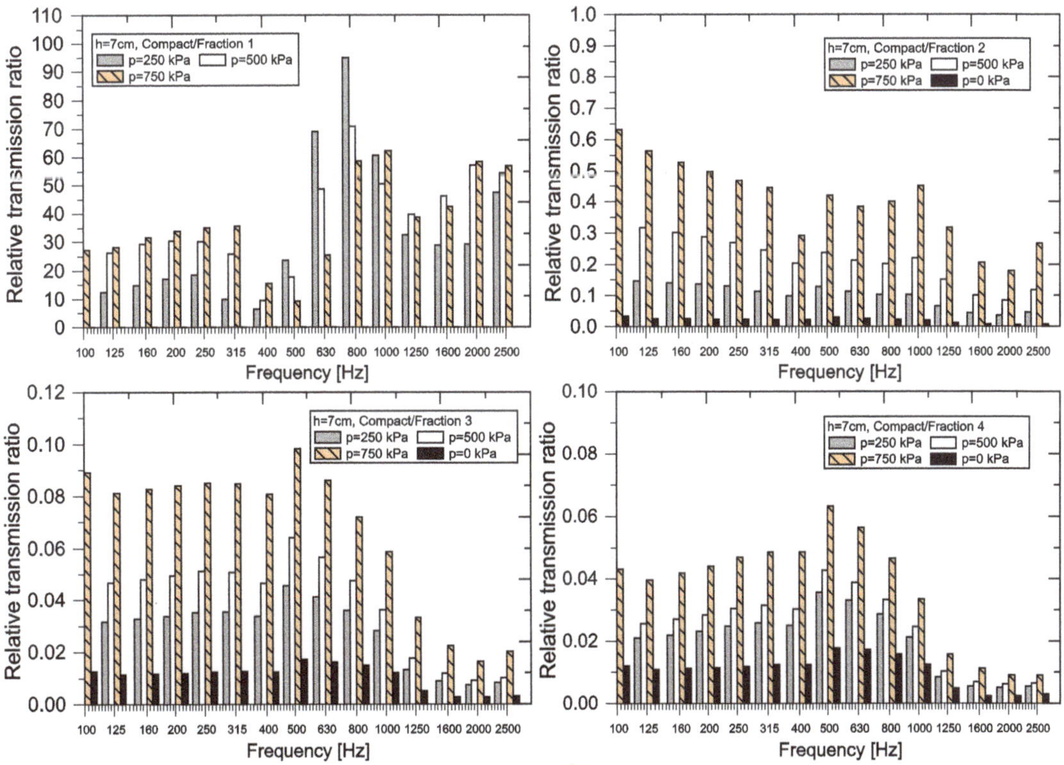

Figure 14. Graphical representation of RTR for RRC samples (h = 7 cm).

A graphical representation of the relative transmission ratio value for samples with a thickness of 7 cm is shown in Figure 14.

When we compare compact and loose granular recycled rubber crumb (RRC) of *Fraction 1* (h = 7 cm), we find that in all cases except for the sample with a pressure of p = 0 kPa, the granular RRC dampens noise much more effectively than the *Compact RRC*. In these cases, the values of the relative transmission ratio (RRT) are several times higher than 1, which means that the granular sample transmits significantly less sound. On the other hand, at zero pressure (loose granular crumb, p = 0 kPa), the *Compact RRC* is better, which in this case dampens the sound more effectively—RTR values are in the range of 0.05 to 0.1. Because of these extremely low values, the results for Fraction 1 at p = 0 kPa are not displayed in Figure 14.

For *Fractions 2–4*, the *Compact RRC* is always more effective in noise attenuation than a granular RRC, as the RTR values are always less than 1. Thus, the results confirm that *Fraction 1* granular RRC (p = 250–750 kPa) exhibits better acoustic properties than *Compact RRC* at all thicknesses.

3.3. Model Development via Regression Analysis

When analyzing the effect of selected input variables (pressure p, thickness h, frequency *Freq*) on the output variable transmission loss (TL, dB) of granular recycled rubber crumb, we will proceed using the following model:

$$TL = f(p, h, Freq), \tag{9}$$

or for a compact recycled rubber crumb model:

$$TL = f(h, Freq). \tag{10}$$

The best regression model for *Fraction 1* and *Fraction 2* is a model of the following form:

$$TL_{granular} = \beta_0 + \beta_1 \cdot p + \beta_2 \cdot h + \beta_3 \cdot Freq + \beta_4 \cdot p \cdot h + \varepsilon, \tag{11}$$

or for *Fraction 3* and *Fraction 4*:

$$TL_{granular} = \beta_0 + \beta_1 \cdot p + \beta_2 \cdot h + \beta_3 \cdot Freq + \varepsilon, \tag{12}$$

where β_0, β_i (i = 1,2,3,4) are the parameters of the model; p is the pressure (kPa); h is the thickness of the sample (cm); *Freq* is the frequency (Hz); and $p \times h$ represents the interaction between thickness and pressure.

In the case of compact recycled rubber crumb, the best regression model has the following form:

$$TL_{compact} = \beta_0 + \beta_1 \cdot h + \beta_2 \cdot Freq + \varepsilon. \tag{13}$$

Point estimates of the regression model parameters, standardized coefficients, and their statistical significance (p-value) are summarized in Table 6 (see Equations (9)–(13)). The table also reports the multiple coefficient of determination (R^2) and the overall model significance (p-value). All parameters and the regression models themselves are statistically significant (p-value < α), with the significance level set at $\alpha = 0.05$.

The regression analysis showed that all input variables (pressure p, thickness h, frequency *Freq*, interaction $p \times h$) significantly affect the transmission loss with each investigated fraction. The strength of this relationship between the transmission loss and the input variables is expressed by a multiple coefficient of determination R^2. In the case of *Fraction 1*, the coefficient of determination is equal to 0.82, which means that 82.0% of the variability of the variable transmission loss can be explained by the proposed regression model. Other models have similarly high values (*Fraction 2*: R^2 = 0.955 (95.5%), *Fraction 3*: R^2 = 0.833 (83.8%), *Fraction 4*: R^2 = 0.808 (80.8%), *Compact sample*: R^2 = 0.862 (86.2%)), which represents a strong degree of dependence of the transmission loss variable on the input variables.

These results point to a strong degree of dependence of the transmission loss on the input variables and confirm that the proposed regression models are suitable for predicting the acoustic properties of recycled rubber crumb under different conditions.

Table 6. Regression models—transmission loss (α = 0.05).

Parameters	Coefficients	Standardized Coefficients Beta	*t*-Stat	*p*-Value
Granular recycled rubber crumb—*Fraction 1* (*p*-value < 2×10^{-16}), R^2 = 0.820				
Intercept	1.221	-5.96×10^{-18}	0.716	0.0371
p	1.902×10^{-2}	0.436	5.384	2×10^{-7}
h	2.437×10^{2}	0.408	7.271	$<10^{-10}$
Freq	4.158×10^{-3}	0.219	6.561	$<10^{-10}$
p × h	2.333×10^{-1}	0.300	3.237	1.1×10^{-3}
Granular recycled rubber crumb—*Fraction 2* (*p*-value < 2×10^{-16}), R^2 = 0.955				
Intercept	2.755	3.9×10^{-16}	7.822	$<10^{-16}$
p	1.562×10^{-3}	0.086	2.156	3.10^{-2}
h	93.09	0.377	13.315	$<10^{-16}$
Freq	1.408×10^{-3}	0.189	11.412	$<10^{-13}$
p × h	2.115×10^{-1}	0.656	14.154	$<10^{-13}$
Granular recycled rubber crumb—*Fraction 3* (*p*-value < 2×10^{-16}), R^2 = 0.838				
Intercept	2.482×10^{-1}	3.9×10^{-16}	0.883	0.379
p	6.643×10^{-3}	0.579	18.954	$<10^{-16}$
h	91.69	0.609	18.930	$<10^{-16}$
Freq	1.417×10^{-3}	0.324	9.514	$<10^{-16}$
Granular recycled rubber crumb—*Fraction 4* (*p*-value < 2×10^{-16}), R^2 = 0.808				
Intercept	5.317×10^{-1}	3.8×10^{-16}	2.109	0.036
p	3.809×10^{-3}	0.415	11.921	$<10^{-16}$
h	76.3	0.618	17.748	$<10^{-16}$
Freq	1.75×10^{-3}	0.466	13.387	$<10^{-16}$
Compact Sample (*p*-value < 2×10^{-16}), R^2 = 0.862				
Intercept	2.75×10^{-2}	2.2×10^{-16}	0.024	0.981
h	3.12	0.923	3.171	$<10^{-16}$
Freq	2.43×10^{-3}	0.191	15.251	3×10^{-3}

The values of the parameters obtained by the regression models show that all input variables have a positive effect on the value of transmission loss. The analysis of standardized Beta coefficients (see the Beta column in Table 6) shows the following:

1. Fraction 1: the pressure variable (p) has the most significant effect on transmission loss, with the thickness (h) of the samples having a slightly smaller effect.
2. Fraction 2: the interaction between the variable h and the variable p ($p \times h$) has the most significant influence.
3. Fraction 3: the variables pressure (p) and thickness (h) have a comparable effect on transmission loss.
4. Fraction 4: the thickness variable (h) has the most significant effect on transmission loss; the other two variables have a comparable effect.
5. Compact recyclable crumb: variable thickness (h) has by far the most significant influence on transmission loss.

These results confirm that optimizing acoustic properties depends on the appropriate selection and combination of input parameters.

4. Discussion and Conclusions

This paper evaluated the acoustic properties of recycled rubber crumb samples, with the main criterion being the transmission loss (TL, dB) measured by an impedance tube in accordance with international standards. The samples were prepared in two basic forms: loose granular (p = 0 kPa) and compacted under pressure (250 kPa, 500 kPa, 750 kPa). For comparison, a compact sample cut from a commercially available panel in three thicknesses (2 cm, 4.5 cm, 7 cm) was also analysed.

Based on experimental measurement of the transmission coefficients τ_f (in %) and the average transmission loss TL_{avg} (in dB) for granular and compact recycled rubber crumb (RRC), the following key relationships between pressure p, thickness h, and frequency response *Freq* were identified:

1. Scientific and professional studies by the authors [15,26,37] focus on the analysis of the structure of individual materials and its influence on acoustic properties. The results of the research show that *Fraction 1* has the best soundproofing properties, which is reflected in the lowest value of the transmission coefficient. This trend is in line with the law of mass, according to which smaller particles form a more compact structure with a higher density, resulting in more effective sound damping. The authors [26,37] state in their publications that the size of the fraction and compactness of materials significantly affect acoustic properties, while smaller and more compact particles improve the ability of the material to dampen the propagation of sound waves. This knowledge provides a theoretical basis for optimizing the composition and structure of materials in order to achieve a higher level of sound insulation.
2. The results of experimental measurements show that thicker samples of material have much better sound insulation properties than thinner samples. This phenomenon is due to the fact that with a greater thickness of material, several factors are better applied together—in particular, the higher mass, the ability of the material to absorb vibrations (viscoelastic damping), and a more complex internal structure. These findings are also confirmed by other authors [3,4], who emphasize that it is the combination of a higher specific gravity and a more complex structure that is key to the best possible sound insulation.
3. Experimental results show that with increasing pressure, the porosity of the material decreases, and its density increases. These changes lead to a decrease in the value of the transmission coefficient τ_f in all the investigated fractions, which means that the material better prevents the attenuation of sound. The most pronounced effect was observed in *Fraction 1*, where the difference in attenuation between the loose granular material and the sample exposed to pressure (p = 750 kPa) was very significant. This clearly confirms that mechanical compression, i.e., compaction of the material, significantly improves its acoustic properties. This positive effect is mainly due to the fact that compression eliminates air gaps, creating a more homogeneous and dense material structure that better absorbs and dissipates sound energy.
4. The analysis of the transmission coefficients of recycled rubber crumb shows that this material has significant potential as an effective sound insulation element, especially when using fine fractions exposed to pressure. The frequency dependence of the transmission coefficients corresponds to the theoretical assumptions of acoustics of porous materials, which is also confirmed by several scientific studies [3,15,23]. These results suggest that recycled rubber crumb may be a suitable alternative for applications aimed at improving sound insulation.
5. The analysis confirmed that the acoustic properties of recycled rubber crumb are significantly influenced by the synergistic action of the mass law, structural heterogeneity, and resonance phenomena. The increase in the thickness of the samples caused the

critical frequency to shift to a lower frequency band, which has a significant impact on the effectiveness of sound insulation. For practical use, it is recommended to combine fractions of different sizes and apply a layered structure of the material, which will achieve more effective suppression of broad-spectrum noise.

If it is necessary to replace the compact recycled rubber crumb with another fraction with similar or better transmission loss values, we recommend considering the following options and approaches:

1. Fraction 3 and Fraction 4: these fractions are not recommended for use if the goal is to achieve better transmission loss, as their acoustic insulation properties are insufficient.
2. Fraction 1: Loose granular RRC (at a pressure of 0 kPa) is not suitable for use, as it does not provide sufficient attenuation. On the contrary, it is recommended to apply a pressure of 500 kPa or 700 kPa, which will significantly improve the acoustic properties of this fraction.
3. Fraction 2: Loose granular RRC (0 kPa) or granular RRC with a pressure of 250 kPa is also not suitable, because the attenuation is low in these cases. The use of a pressure of 500 kPa or 700 kPa is recommended, with a greater sample thickness further improving the transmission loss values for this fraction.

The results of the regression analysis clearly indicate a statistically significant influence of the analyzed input parameters (pressure, thickness, frequency) on the magnitude of the transmission loss. The developed regression models show a high degree of statistical significance and represent a suitable tool for predictive modeling of transmission loss behavior depending on variable input parameters for each type of sample (loose granular or compacted recycled rubber crumb).

The experimental data and results from this study have clear practical value. The findings, together with the regression models we developed, can guide the design of acoustic panels made from recycled materials, especially when estimating sound absorption performance as functions of layer thickness, particle size, and other parameters. These insights also provide a solid foundation for future research: by incorporating additional variables, subsequent models could further advance sustainable, high-performance acoustic solutions.

The morphology and particle size of the rubber crumb significantly affect its acoustic performance. Fine fractions, when highly compacted, provide a versatile solution across a broad frequency spectrum, whereas coarser fractions are better suited for targeted applications in the low-frequency range. The irregular shape and rough surface of the recycled rubber particles create variable inter-particle gaps during compaction, yielding a heterogeneous microstructure that influences transmission loss. As a result, compaction density varies: fine fractions ($\leq$1 mm) form a relatively uniform matrix with small, evenly distributed pores, whereas coarser fractions produce larger, irregular voids. In addition, size variability within each fraction introduces localized density fluctuations, which partly account for the statistical scatter observed in transmission loss measurements.

At the same time, the experimental results show that with the right choice of fraction and applied pressure, it is possible to achieve optimal acoustic properties of recycled rubber crumb for specific applications. This underlines its potential as an environmentally friendly material for sound insulation. Recycled rubber is a sustainable alternative to traditional materials such as mineral wool, with the added advantage of being highly resistant to moisture, chemicals, and similar factors. This feature expands its potential uses, especially in industrial environments. The main advantage of loose recycled rubber materials lies in their favorable acoustic properties and environmental compatibility, as they do not contain any added chemical binders. A potential drawback, however, may be the technological challenges associated with applying these materials in noise barrier wall constructions. The authors of [38] state that the acoustic properties of materials based on recycled tires are comparable to those of standard commercially available materials used in noise barrier walls. Several authors [6,25,29] highlight the suitability of recycled materials for soundproofing applications and also point out their environmental benefits. Recycled rubber crumb thus combines technical efficiency with environmental sustainability and represents a forward-looking solution for modern acoustic applications.

5. Patents

The test cartridge described in this paper is the subject of patent No. 289324, "Test Cartridge for Expanding the Possibilities of Measuring Bulk Materials in an Impedance Tube", which was granted by the Industrial Property Office of the Slovak Republic.

The laboratory device for filling and compaction with digital control described in this article is the subject of published patent application no. 50022-2022, "Device for Measured Filling of Test Cartridges with Bulk Materials", which was granted by the Industrial Property Office of the Slovak Republic.

Author Contributions: Conceptualization, M.B., M.A., M.P. and M.M.; methodology, M.A., M.P. and M.M.; software, M.A. and M.M.; validation, M.B., M.A., M.P. and M.M.; formal analysis, M.A. and M.P.; investigation, M.A., M.P. and M.M.; resources, M.B., M.A. and M.P.; data curation, M.B. and M.M.; writing—original draft preparation, M.B., M.A., M.P. and M.M.; writing—review and editing, M.A., M.P. and M.M.; visualization, M.A. and M.P.; supervision, M.B.; project administration, M.P. and M.M.; funding acquisition, M.B. All authors have read and agreed to the published version of the manuscript.

Funding: This research received no external funding.

Institutional Review Board Statement: Not applicable.

Informed Consent Statement: Not applicable.

Data Availability Statement: The original contributions presented in the study are included in the article. Further inquiries can be directed to the corresponding author.

Acknowledgments: This work was supported by the projects of the Scientific Grant Agency of the Ministry of Education, Science, Research and Sport of the Slovak Republic, project KEGA No. 032TUKE-4/2025 and project UNIVNET No. 0201/0082/19.

Conflicts of Interest: The authors declare no conflicts of interest.

References

1. Niresh, J.; Neelakrishnan, S.; Subharani, S.; Kannaian, T.; Prabhakaran, R. Review of acoustic characteristics of materials using impedance tube. *ARPN J. Eng. Appl. Sci.* **2015**, *10*, 3319–3326.
2. Arunkumar, B.; Jeyanthi, S. Design and analysis of impedance tube for sound absorption measurement. *ARPN J. Eng. Appl. Sci.* **2017**, *12*, 1400–1405.
3. Desarnaulds, V.; Costanzo, E.; Carvalho, A.P.O.; Arlaud, B. Sustainability of acoustic materials and acoustic characterization of sustainable materials. In Proceedings of the 12th International Congress on Sound and Vibration (ICSV12), Lisbon, Portugal, 11–14 July 2005.
4. Bhingare, N.H.; Prakash, S.; Jatti, V.S. A review on natural and waste material composite as acoustic material. *Polym. Test.* **2019**, *80*, 106142. [CrossRef]
5. Berardi, U.; Iannace, G. Acoustic characterization of natural fibers for sound absorption applications. *Build. Environ.* **2015**, *94*, 840–852. [CrossRef]
6. Ashby, M.F. *Materials and the Environment*, 3rd ed.; Butterworth-Heinemann: Oxford, UK, 2021; pp. 267–294.
7. McGinnes, C.; Kleiner, M.; Xiang, N. An environmental and economical solution to sound absorption using straw. *J. Acoust. Soc. Am.* **2005**, *118*, 1869. [CrossRef]
8. D'Alessandro, F.; Schiavoni, S.; Bianchi, F. Straw as an acoustic material. In Proceedings of the 24th International Congress on Sound & Vibration, London, UK, 23–27 July 2017.

9. Colom, X.; Girbau, J.; Marin, M.; Formela, K.; Saeb, R.M.; Carrillo, F.; Cañavate, J. New approach in the reuse of modified ground tire rubber as thermal and acoustic insulation to be used in civil engineering. *J. Mater. Cycles Waste Manag.* **2023**, *25*, 3557–3566. [CrossRef]
10. Sambucci, M.; Marini, D.; Sibai, A.; Valente, M. Preliminary mechanical analysis of rubber-cement composites suitable for additive process construction. *J. Compos. Sci.* **2020**, *4*, 120. [CrossRef]
11. Sambucci, M.; Valente, M. Ground waste tire rubber as a total replacement of natural aggregates in concrete mixes: Application for lightweight paving blocks. *Materials* **2021**, *14*, 7493. [CrossRef]
12. Sambucci, M.; Marini, D.; Valente, M. Tire recycled rubber for more eco-sustainable advanced cementitious aggregate. *Recycling* **2020**, *5*, 11. [CrossRef]
13. Akiwate, D.; Mohan, S.R.; Venkatesham, B.; Suryakumar, S. Acoustic characterization of additive manufactured layered porous materials. In Proceedings of the 48th International Congress and Exhibition on Noise Control Engineering (INTERNOISE 2019), Madrid, Spain, 16–19 June 2019.
14. Labašová, E.; Ďuriš, R. Measurement of the Acoustic Absorption Coefficient by Impedance Tube. *Res. Pap. Fac. Mater. Sci. Technol. Slovak Univ. Technol.* **2019**, *27*, 94–101. [CrossRef]
15. Sandu, C.; Totu, A.G.; Trifu, A.T.; Deaconu, M. Advances in Experimental Research on the Influence of Friction Powders on Acoustic Liners (Helmholtz Resonators). *Aerospace* **2023**, *10*, 1000. [CrossRef]
16. Vasile, O.; Iaşnicu, I.; Iatan, R. The analysis of sound absorbing performances for composite plates containing recycled textile wastes. *U.P.B. Sci. Bull. Ser. D* **2016**, *78*, 213–220.
17. Djikou, M.S.; Houngan, C.A.; Anjorin, M.; Vianou, A.; Degan, G. Measurement of the acoustic absorption coefficient of some local building materials for residential buildings using the kundt tube method. In Proceedings of the 33rd International Conference on Vibroengineering (Procedia), Zittau, Germany, 24–26 September 2018.
18. Suhanek, M.; Jambrosic, K.; Domitrovic, H. Student project of building an impedance tube. *J. Acoust. Soc. Am.* **2008**, *123*, 3616. [CrossRef]
19. Pierre, J.; Guillermic, R.-M.; Elias, F.; Drenckhan, W.; Leroy, V. Acoustic characterisation of liquid foams with an impedance tube. *Eur. Phys. J. E* **2013**, *36*, 113. [CrossRef]
20. Koruk, H. An assessment of the performance of impedance tube method. *Noise Control Eng. J.* **2014**, *62*, 264–274. [CrossRef]
21. Tan, W.H.; Ahmad, R.; Zunaidi, N.H.; Daud, R.; Cheng, E.M. Development of an Indigenous Impedance Tube. *Appl. Mech. Mater.* **2015**, *786*, 149–155. [CrossRef]
22. Neto, B.; Pereira, I.; Futatsugi, S.; Mareze, P.; Brandão, E.; Fonseca, W. Experimental analysis of the dispersion in the measurement of the absorption coefficient with the impedance tube. In Proceedings of the 47th International Congress and Exposition on Noise Control Engineering (INTERNOISE 2018), Chicago, IL, USA, 26–29 August 2018.
23. Ipek, O.; Koru, M.; Serce, O.; Karabulut, H.; Hıdıroglu, M. Determining sound absorbing and transmission loss properties of rubbers used in automotive industries. *Eurasia Proc. Sci. Technol. Eng. Math.* **2018**, *4*, 137–148.
24. Designing Acoustics in Industry. Available online: https://www.isover-technical-insulation.com/ (accessed on 13 May 2025).
25. Ivanova, Y.; Partalin, T. Experimental study of acoustical properties of recycled rubber panels. *Eurasia Proc. Sci. Technol. Eng. Math.* **2024**, *32*, 379–388. [CrossRef]
26. Canli, E.; Sertkaya, A.A.; Canli, E.; Sarikaya, A.; Güler, C. Determination of Sound Absorption Coefficient and Transmission Loss Properties of Almond Shell Particle Boards. In Proceedings of the 5th International Conference on Materials Science, Mechanical and Automotive Engineering and Technology (IMSMATEC'22), Nevşehir, Türkiye, 2–4 September 2022.
27. 2023 ESG Report Press Release. Available online: https://libertytire.com/Resources/2023-ESG-Report-Press-Release/ (accessed on 13 May 2025).
28. Bušić, R.; Miličević, I.; Šipoš, T.K.; Strukar, K. Recycled Rubber as an Aggregate Replacement in Self-Compacting Concrete—Literature Overview. *Materials* **2018**, *11*, 1729. [CrossRef]
29. Medina, N.F.; Garcia, R.; Hajirasouliha, I.; Pilakoutas, K.; Guadagnini, M.; Raffoul, S. Composites with recycled rubber aggregates: Properties and opportunities in construction. *Constr. Build. Mater.* **2018**, *188*, 884–897. [CrossRef]
30. Osvaldová, M.; Potkány, M. Economic principles and possibilities of using recycled material based on wood—Plastic—Rubber. In Proceedings of the 15th International Scientific Conference WoodEMA 2022, Trnava, Slovakia, 8–10 June 2022.
31. Abdelmonem, A.; El-Feky, M.S.; Nasr, E.-S.A.R.; Kohail, M. Performance of high strength concrete containing recycled rubber. *Constr. Build. Mater.* **2019**, *227*, 116660. [CrossRef]
32. Badida, M.; Moravec, M.; Pinosova, M.; Andrejiova, M.; Pástor, K.; Nováková, A.; Dzuro, T. Analysis and Research on the Use of Bulk Recycled Materials for Sound Insulation Applications. *Sustainability* **2022**, *14*, 11539. [CrossRef]
33. *STN EN ISO 10534-2 (730537)*; Acoustics—Determination of Sound Absorption Coefficient and Impedance in Impedances Tubes—Part 2: Transfer-Function Method (ISO 10534-2:1998). International Organization for Standardization: Geneva, Switzerland, 2024.
34. Jedidi, M.; Boulila, A.; Benjeddou, O.; Soussi, C. Crumb Rubber Effect on Acoustic Properties of Self- Consolidating Concrete. *Int. J. Therm. Environ. Eng.* **2014**, *8*, 69–76. [CrossRef]

35. Gumanova, V.; Sobotova, L.; Dzuro, T.; Badida, M.; Moravec, M. Experimental Survey of the Sound Absorption Performance of Natural Fibres in Comparison with Conventional Insulating Materials. *Sustainability* **2022**, *14*, 4258. [CrossRef]
36. Maděra, J.; Fořt, J.; Fiala, L.; Kočí, V.; Černý, R. Acoustic Properties of Rubber Crumb Boards Originated from End-of-Life Tires. In Proceedings of the International Conference of Numerical Analysis and Applied Mathematics ICNAAM 2021, Rhodes, Greece, 20–26 September 2021.
37. Maderuelo-Sanz, R.; Morillas, J.; Miguel, B.; Martin-Castizo, M.; Escobar, V.G.; Gozalo, G.R. Acoustical performance of porous absorber made from recycled rubber and polyurethane resin. *Lat. Am. J. Solids Struct.* **2013**, *10*, 585–600. [CrossRef]
38. Lu, K.; Shang, X.; Yuan, M. Preparation and performance analysis of porous materials for road noise abatement using waste rubber tires. *Appl. Acoust.* **2025**, *228*, 110374. [CrossRef]

materials

MDPI

Article

Theoretical Analysis of Low-Frequency Sound Absorption Owing to the Vibration of Lightweight Powder Using a 1D Beam Model

Shuichi Sakamoto [1,*], Yuya Kawakami [2], Hiroaki Soeta [2] and Yosuke Kubo [2]

[1] Department of Engineering, Niigata University, Ikarashi 2-no-cho 8050, Nishi-ku, Niigata 950-2181, Japan
[2] Graduate School of Science and Technology, Niigata University, Ikarashi 2-no-cho 8050, Nishi-ku, Niigata 950-2181, Japan
* Correspondence: sakamoto@eng.niigata-u.ac.jp; Tel.: +81-25-262-7003

Abstract: Lightweight powder-based sound-absorbing materials are characterized by sound absorption peaks at lower frequencies compared to other sound absorption materials of the same thickness. This behavior is attributed to the excitation of longitudinal vibration modes in the powder particles by incident sound waves, wherein acoustic energy is converted into kinetic energy and subsequently dissipated through interparticle interactions. These lightweight, fine powders are artificially engineered acoustic materials. Despite their structural simplicity, they exhibit emergent and complex sound absorption behaviors through fundamental vibrational mechanisms. Representing the powder layer with a transfer matrix simplifies model-based development and enhances versatility as an acoustic element. The powder layer was modeled as a longitudinally oscillating 1D beam, and transfer matrix of the powder layer was derived. To verify the obtained transfer matrix, the experimental values were compared with the theoretical values for a single powder layer. In addition, both were compared for the case of other acoustic elements stacked on top of each other, which were close to each other. The theoretical values were compared with the experimental values, which were close to each other.

Keywords: sound absorption coefficient; lightweight powder; longitudinal vibration; 1D beam model; transfer matrix

1. Introduction

Lightweight powder-based sound-absorbing materials (SAMs) are characterized by sound absorption peaks at lower frequencies compared to other SAMs of the same thickness. Therefore, powder-based SAMs are considered enablers of smaller, lighter SAM designs.

Lightweight powders with particle sizes of several tens of micrometers exhibit sound absorption peaks at lower frequencies than other porous SAMs. In such powders, boundary layer viscosity—typically the dominant absorption mechanism in porous materials—has minimal influence. Instead, acoustic energy is attenuated through interparticle interactions induced by longitudinal vibrations of the powder particles, excited by incident sound waves [1]. These lightweight, fine powders are artificially engineered acoustic materials. Despite their structural simplicity, they exhibit emergent and complex sound absorption behaviors through fundamental vibrational mechanisms [2].

The lightweight powders analyzed in this study have bulk densities in the range of 13.9–82 kg/m^3, comparable to that of conventional sound-absorbing materials such as glass wool, which typically ranges from 10 to 96 kg/m^3. At a layer thickness of 20 mm, the sound absorption peak of glass wool occurs around 2000 Hz, while that of the lightweight powders lies between 500 and 1000 Hz. Therefore, these powders demonstrate superior absorption in the low-frequency range compared to standard porous materials. Regarding acoustic powder properties, studies exist on the sound absorption properties of powder beds made of binary powder mixtures [3], modal analysis of powder layers as a five-degree-of-freedom forced vibration system and calculation of acoustic impedance [4], and exploration of the sound absorption effect of boundary-layer viscosity beyond sound absorption by vibration [5]. Studies have also examined the conditions under which the sound absorption effect of powder layer vibration appears [6,7] and the vibration isolation and sound insulation properties [8] of silica aerogels. In recent years, the Biot model [9] has been used to study the impact of casing on the sound absorption of fine spherical-granular materials [10].

SAMs are often combined with several other materials to achieve a higher sound absorption effect. The transfer matrix method is a useful theoretical tool for analyzing SAMs combined in this manner. Therefore, representing the powder layer with a transfer matrix simplifies model-based development and enhances versatility as an acoustic element. Previous studies' theoretical analyses [4,5] have enabled powder layer acoustic impedance determination. However, the absence of transfer matrices prevented theoretical value calculations for combinations with other SAMs. A powder layer transfer matrix was attempted using the two-thickness method [11], but the theoretical values deviated from the experimental values in the high-frequency range.

In this study, the powder layer was modeled as a longitudinally oscillating 1D beam, and transfer matrix of the powder layer was derived. The Biot model [9], commonly used to derive sound absorption coefficients, requires nine experimentally determined parameters. In contrast, the model proposed in this study allows for theoretical analysis based on only three of these nine parameters: bulk density, loss factor, and a newly introduced parameter, the first-order sound absorption peak frequency. This reduction in required parameters represents a significant simplification. The loss factors required in the theoretical analysis were also determined. To verify the obtained transfer matrix, the experimental values were compared with the theoretical values for a single powder layer. In addition, both were compared for the case of other acoustic elements stacked on top of each other. This method is anticipated to be applicable to the design and improvement of acoustic performance in systems such as loudspeaker enclosures incorporating fine powder materials [12].

2. Measurement Overview and Preliminary Investigation

2.1. Measurement Apparatus and Samples

Figure 1 shows a schematic of the actual measurement apparatus for measuring the normal incident sound absorption coefficient (SAC). The measurement sample was mounted in a Brüel–Kjær Type 4206 two-microphone impedance-measuring tube (Nærum, Denmark).

The SAC was measured as follows: A sinusoidal signal was output from a signal generator built into an FFT analyzer DS–3000 (Ono Sokki, Kanagawa, Japan). The sound waves can be treated as plane waves because their wavelength is sufficiently smaller than the impedance tube diameter. The transfer function between the sound pressure signals obtained from the two microphones attached to the impedance tube is then measured by the FFT analyzer. The distance between the microphones is 20 mm. The measured transfer function was used to calculate the normal incident SAC according to ISO 10534-2 (2023) [13]. The inner diameter of the acoustic tube determines the frequency limit for the formation of plane waves, and a small tube with an inner diameter of 29 mm was used in this study. The upper frequency limit that can be measured in this case is 6400 Hz according to ISO 10534-2 [13]. The lower frequency limit is 500 Hz, the measurement accuracy of which is guaranteed by the impedance tube specifications. In this study, the experimental results for 100–6400 Hz are described for reference.

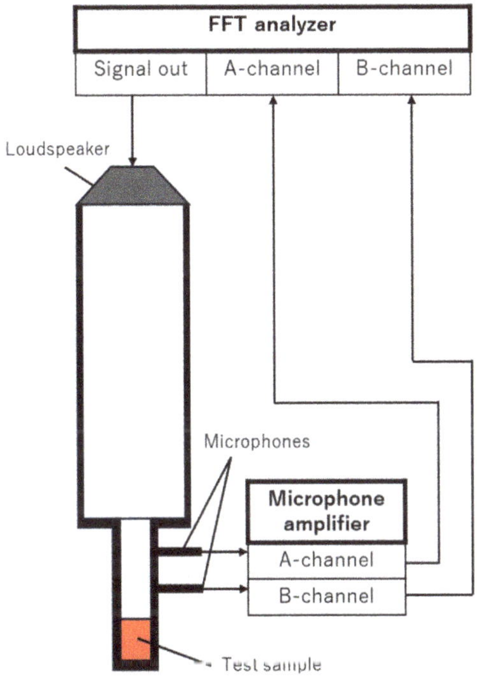

Figure 1. Experimental setup for measuring SACs of powder samples.

The powder controlled in a storage area at a constant temperature (20 °C) and humidity (35%) was filled into an aluminum alloy sample holder (inner diameter 29 mm), and the SAC was measured. The powder was filled by tapping. The powders employed in this study are granular materials. Unlike foam- or fiber-based materials, their volume does not change appreciably under compression. As a result, bulk density remains nearly constant even after tapping, and the corresponding change in sound absorption performance is minimal. Preliminary tests revealed that the sample holder did not affect sound absorption or acoustic properties.

Figure 2 presents the micrographs of the powder samples used for the measurements, while Table 1 summarizes their characteristics. The powders examined in this study are those with bulk densities classified as exhibiting significant sound absorption owing to longitudinal vibration, as indicated in a previous report [7], consistent with the primary objective of this method. Figure 3 shows the sound absorption coefficients of granulated silica (grain diameter: 0.0428 mm; bulk density: 0.0568 g/cm^3) and solid glass beads (grain diameter: 0.05 mm; bulk density: 1.45 g/cm^3). As shown in Figure 3, despite having nearly the same grain size, sound absorption in the heavier powder is dominated by boundary layer viscosity, and minimal absorption occurs in the low-frequency range due to particle vibration. Accordingly, such heavy powders are excluded from the scope of this study.

The particle size represents the average Feret diameter [14] (constant tangential diameter) derived from the micrographs to minimize the influence of particle orientation during measurement. The particle size was not used as a parameter in this study, although it was included as a reference. The first-order peak sound absorption frequencies were obtained from experiments measuring the SAC at a powder layer of 20 mm.

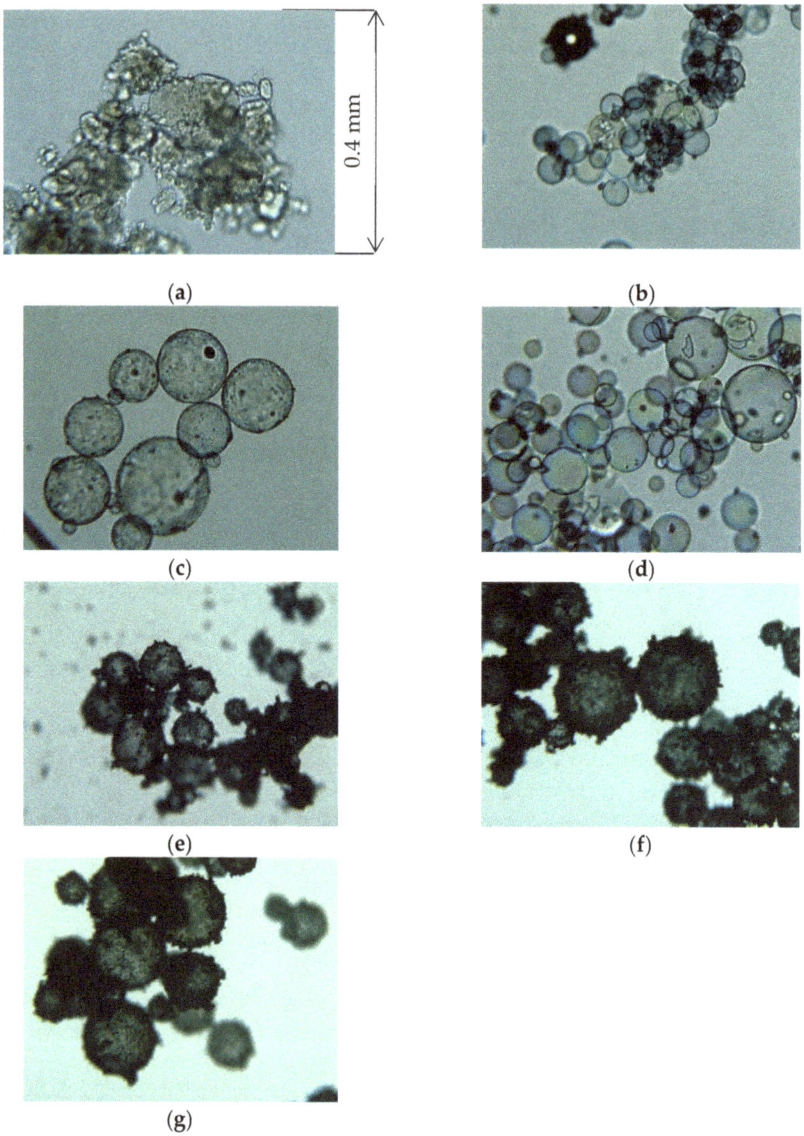

Figure 2. Micrographs of the powder materials: (**a**) granulated silica; (**b**) hollow plastic beads A (Expancel® 920DE40d30); (**c**) hollow plastic beads B (Matsumoto micro sphere® F–80DE); (**d**) hollow glass beads; (**e**) calcium-carbonate-coated hollow plastic beads A (EMC–40(B)); (**f**) calcium-carbonate-coated hollow plastic beads B (EMC–80(B)); (**g**) calcium-carbonate-coated hollow plastic beads C (EMC–120(α)).

Figure 4a shows a schematic diagram of the powder layer alone, and Figure 4b shows a schematic diagram of the powder layer filled on top of the nonwoven fabric, with the nonwoven fabric placed at the upper end of the back air space. The nonwoven fabric used, 3A01A (Toyobo Co., Ltd., Osaka, Japan), is relatively thin and designed with low ventilation resistance to avoid impacting the powder's acoustic properties. The measured ventilation resistance was 0.34×10^3 Pa-s/m and the thickness was 0.39 mm. The ventilation resistance of the nonwoven fabric was measured using a ventilation resistance tester KES-F8-AP1 (Kato Tech Co., Ltd., Kyoto, Japan).

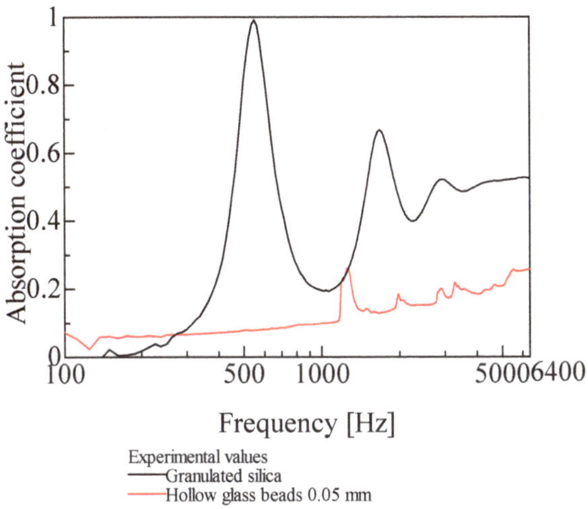

Figure 3. Comparison of sound absorption coefficient between granulated silica and hollow glass beads 0.05 mm.

Table 1. Materials used in the experiments.

	Bulk Density ρ [g/cm^3]	First Peak Frequency f_{p1} [Hz]	Feret Diameter [μm]
Granulated silica (Featherfield Co., Ltd., Hiroshima, Japan)	0.0568	550	43
Hollow plastic beads A (Expancel® 920DE40d30) (Japan Fillite Co., Ltd., Osaka, Japan)	0.0298	1100	42
Hollow plastic beads B (matsumoto micro sphere® F–80DE) (Matsumoto Yushi-Seiyaku Co., Ltd., Osaka, Japan)	0.0139	1837.5	110
Hollow glass beads (Hobbico, Inc., Champaign, IL, USA)	0.0800	950	29
Calcium-carbonate-coated hollow plastic beads A (EMC–40(B)) (Japan Fillite Co., Ltd., Osaka, Japan)	0.082	850	53
Calcium-carbonate-coated hollow plastic beads B (EMC–80(B)) (Japan Fillite Co., Ltd., Osaka, Japan)	0.0707	1012.5	70
Calcium-carbonate-coated hollow plastic beads C (EMC–120(α)) (Japan Fillite Co., Ltd., Osaka, Japan)	0.0628	1137.5	120

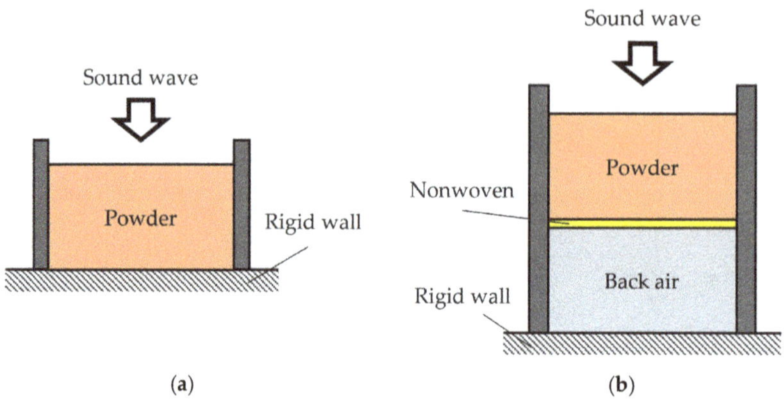

Figure 4. Schematic of the samples: (**a**) powder layer only; (**b**) powder layer with back air space.

2.2. Demonstration of Sound Absorption Due to Longitudinal Vibration

The preliminary experimental results are presented to demonstrate sound absorption caused by the longitudinal vibration of a powder layer. These results were confirmed by the authors before the start of this study [4]. They are included in the first part of this report to address potential doubts regarding the existence of longitudinal vibrations.

The method for demonstrating that the powder absorbs sound due to longitudinal vibrations is as follows: If the powder is compressed to prevent longitudinal vibration, the sound-absorbing effect should be significantly reduced. To test this, we measured the sound absorption coefficient of a powder layer with a slightly greater volume than the sample tube. The powder was then pressed down using a nonwoven fabric (3A01A) to inhibit longitudinal vibrations. A schematic representation of this setup is shown in Figure 5.

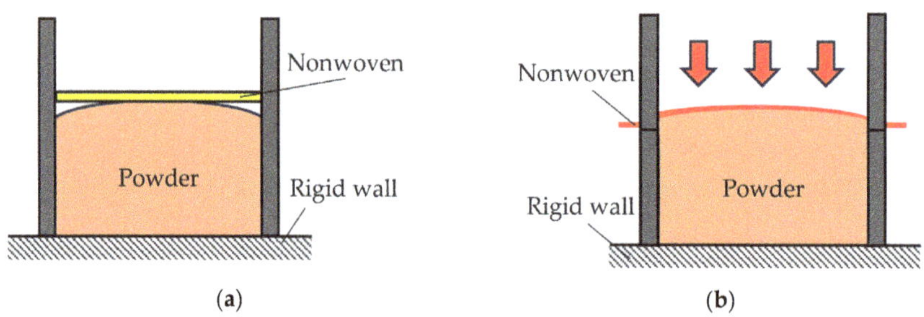

Figure 5. Schematic representation of the samples: (**a**) nonwoven fabric placed on top; (**b**) suppression of longitudinal vibration.

Figure 6a,b presents the experimental results for two different powders: granulated silica and hollow glass beads. When the powders were pressed down to suppress longitudinal vibrations, their sound absorption performance was significantly reduced. This confirms that the sound absorption properties of lightweight powders are primarily due to their longitudinal vibrations.

One possible concern is that the observed change in sound absorption may be attributed to the acoustic effect of the nonwoven fabric rather than the suppression of longitudinal vibrations. To address this, Figure 6a also includes measurements where the nonwoven fabric was simply placed on top of the powder layer without pressing it down. The results show that this had a negligible effect on sound absorption. However, when the fabric was merely placed on the powder layer, the sound absorption peak frequency slightly decreased, and the peak absorption value slightly increased. This is likely due to the slight increase in the powder layer's thickness, as mentioned earlier.

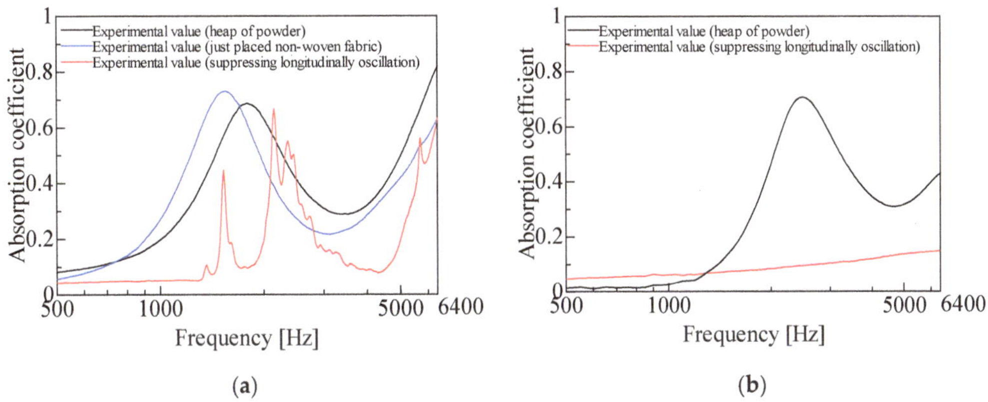

Figure 6. Comparison of experimental values: (**a**) granulated silica, l_p = 5 mm; (**b**) hollow glass beads, l_p = 5 mm.

3. Theoretical Analysis

3.1. Transfer Matrix Method

In this study, the theoretical SAC values were calculated using the transfer matrix method. For the acoustic element shown in Figure 7, if the sound pressure and particle velocity are p_1, p_2 and u_1, u_2 for the incident and terminal ends of the acoustic element, respectively, the acoustic element can be expressed using the transfer matrix as follows:

$$\begin{bmatrix} p_1 \\ Su_1 \end{bmatrix} = \begin{bmatrix} A & B \\ C & D \end{bmatrix} \begin{bmatrix} p_2 \\ Su_2 \end{bmatrix} \quad (1)$$

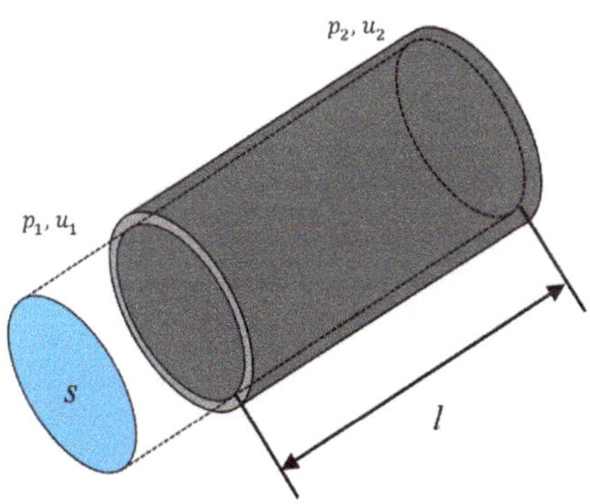

Figure 7. Acoustic element.

Based on the 1D wave equation, the transfer matrix for sound pressure and volume velocity can be expressed as follows:

$$T = \begin{bmatrix} A & B \\ C & D \end{bmatrix} = \begin{bmatrix} cosh(\gamma l) & \frac{Z_c}{S} sinh(\gamma l) \\ \frac{S}{Z_c} sinh(\gamma l) & cosh(\gamma l) \end{bmatrix} \quad (2)$$

where S is the cross-sectional area of the acoustic element, l is the length, γ is the propagation constant, and Z_c is the characteristic impedance. Therefore, to obtain the transfer matrix of the acoustic element, γ and Z_c are required.

3.2. Characteristic Impedance and Propagation Constants of Powder Layers

In previous studies, vibration analysis of longitudinally oscillating beams was conducted using a transfer matrix [15–17]. In the present study, the powder layer was used as a 1D beam model to derive the transfer matrix. In other research domains, analytical models have been extended to two- and three-dimensional representations [18]. However, in acoustic engineering, such extensions are typically considered too complex for analytical approaches and are usually addressed using commercial numerical simulation software (Comsol Multiphysics® Version 5.2). In contrast, the present study seeks a simplified theoretical model that can clearly capture the underlying physical mechanisms.

The powder layer was subjected to incident plane waves normal to the sample surface. Therefore, only powder layer longitudinal vibrations along the tube's longitudinal direction are assumed to occur, with no radial vibrations generated. The peak sound absorption frequencies of the powder layer increase threefold from the first to the second order and fivefold from the first to the third order. This is similar to the vibration modes of a longitudinally vibrating 1D beam with one fixed end. Therefore, the propagation of sound waves in the powder layer is modeled as a longitudinally vibrating 1D beam. Figure 8a presents the five degrees of freedom (5-DOF) model used in previous work, while Figure 8b shows the proposed longitudinally vibrating 1D beam model.

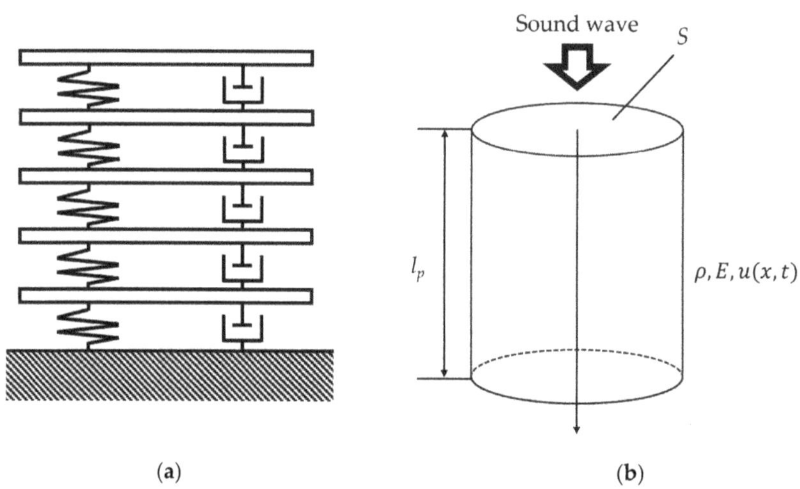

Figure 8. Analytical models: (**a**) 5-DOF; (**b**) longitudinally vibrating 1D beam model.

If the displacement u of the beam is a function of the position x on the x-axis and time t, $u(x, t)$, the wave equation is obtained as follows:

$$\rho \frac{\partial^2 u}{\partial t^2} = E \frac{\partial^2 u}{\partial x^2} \tag{3}$$

where ρ is the beam density, which is equal to the powder layer bulk density. E is the Young's modulus of the beam (=powder layer).

The velocity C of the longitudinal elastic wave propagating in the model is given by Equation (4).

$$C = \sqrt{\frac{E}{\rho}} \tag{4}$$

The first-order eigenfrequency occurs at a frequency such that four times the thickness of the powder layer corresponds to the wavelength. From the relationship between the speed of sound in the medium, frequency, and wavelength, the following equation is obtained:

$$f_{p1} = \frac{C}{4l_p} \tag{5}$$

From Equations (4) and (5), Equation (6) is obtained for Young's modulus of the powder layer.

$$E = \rho(4f_{p1}l_p)^2 \tag{6}$$

The governing equation for powder layer vibration is the wave equation, which is the same as that used for the propagation of sound waves in air. Therefore, as in Equation (2), for a 1D beam mode, the transfer matrix T_p for the pressure and volume velocity at both ends of the beam can be established. The transfer matrix T_p for the powder layer is derived using expressions for the propagation constant and characteristic impedance of the surrounding air layer.

$$T_p = \begin{bmatrix} cosh(\gamma_p l_p) & \frac{Z_p}{S} sinh(\gamma_p l_p) \\ \frac{S}{Z_p} sinh(\gamma_p l_p) & cosh(\gamma_p l_p) \end{bmatrix} \tag{7}$$

$$\gamma_p = \rho C = j\omega\sqrt{\frac{\rho}{E}} \tag{8}$$

$$Z_p = \frac{i\omega}{C} = \sqrt{\rho E} \tag{9}$$

This takes the same form as the transfer matrix of the acoustic elements in Equation (2). Thus, Z_p and γ_p can be treated as the characteristic impedance and propagation constants of the powder layer, respectively.

Equation (7) represents the transfer matrix of the beam without attenuation. Here, energy dissipation occurs in the powder layer due to inter-particle interactions. Therefore, Young's modulus E of the powder obtained in Equation (6) is expressed using the loss factor η as follows:

$$E^* = E(1 + i\eta) \tag{10}$$

E^* is the complex Young's modulus, where the imaginary part represents damping; The loss factor η is a constant. By substituting E^* into E in Equations (8) and (9), the transfer matrix of the beam with damping can be obtained. The loss factor is defined as the value that achieves the best agreement between theoretical and experimental results at the first-order peak frequency, with details provided in the next section.

3.3. Characteristic Impedance and Propagation Constants of Nonwoven Fabrics

The Delany–Bazlry [19], Miki [20], and Komatsu [21] models are models of the airborne sound properties of fabric materials.

In this study, the Miki model [20] was used to calculate the characteristic impedance and propagation constants. The flow resistivity σ of the nonwoven fabric was obtained using Equation (11).

$$\sigma = \frac{R_{non}}{l_{non}} \tag{11}$$

where R_{non} is the nonwoven fabric ventilation resistance and l_{non} is its thickness.

By substituting the flow resistivity σ into the following equation, the propagation constants γ_{non} and characteristic impedance Z_{non} of the nonwoven fabric are obtained, where f is the frequency.

$$Z_{non} = \left(1 + 0.0699\left(\frac{f}{\sigma}\right)^{-0.632}\right) - j\left(-0.107\left(\frac{f}{\sigma}\right)^{-0.632}\right) \quad (12)$$

$$\gamma_{non} = \left(\left(\frac{\omega}{c_0}\right)\left(0.160\left(\frac{f}{\sigma}\right)^{-0.618}\right)\right) + j\left(\left(\frac{\omega}{c_0}\right)\left(1 + 0.109\left(\frac{f}{\sigma}\right)^{-0.618}\right)\right) \quad (13)$$

3.4. Characteristic Impedance and Propagation Constants of the Air Space

The characteristic impedance and propagation constants Z_0 and γ_0 of the air layer can be expressed in terms of $\rho_0 c_0$ and wavenumber, respectively, neglecting attenuation, as follows:

$$Z_0 = \rho_0 c_0 \quad (14)$$

$$\gamma_0 = \frac{i\omega}{c} \quad (15)$$

3.5. Cascade Connection of Four-Terminal Networks in Equivalent Circuits

Having obtained the characteristic impedance and propagation constants for the powder, nonwoven, and air layers in Sections 3.1–3.4, they can be substituted into Equation (1) to obtain the respective transfer matrices. Let T_p, T_{non}, and T_{ba} be the transfer matrices of the powder, nonwoven, and air layers, respectively.

Figure 9a,b shows the equivalent circuits corresponding to Figure 4a,b, respectively. When the four-terminal networks corresponding to several acoustic elements are cascaded, as shown in Figure 9b, the overall transfer matrix T_{all} is obtained by multiplying the transfer matrices of the respective acoustic elements. In other words, the transfer matrix T_{all} of the entire acoustic system shown in Figure 4b can be expressed as follows:

$$T_{all} = T_p \times T_{non} \times T_{ba} \quad (16)$$

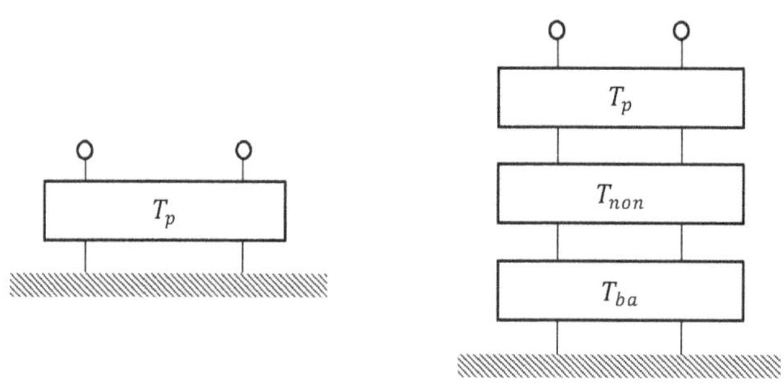

(a) (b)

Figure 9. Equivalent circuit of the samples: (**a**) powder layer only; (**b**) powder layer with back air space.

3.6. SAC Calculation

For the transfer matrix of the acoustic elements shown in Equation (1), the termination of the sample is a rigid wall; so, the particle velocity u_2 = 0 can be assumed to obtain Equation (17).

$$\begin{bmatrix} p_1 \\ Su_1 \end{bmatrix} = \begin{bmatrix} Ap_2 \\ Cp_2 \end{bmatrix} \quad (17)$$

The specific acoustic impedance z_0 of the entire sample viewed internally from the sample incident surface is given as follows:

$$z_0 = \frac{p_1}{u_1} = \frac{A}{C}S \tag{18}$$

where the relationship between the specific acoustic impedance z_0 and the complex reflectance R is expressed in Equation (19).

$$R = \frac{z_0 - \rho_0 c_0}{z_0 + \rho_0 c_0} \tag{19}$$

The overall SAC, α, is expressed by the relationship between the SAC and reflection coefficient R, as shown in Equation (20).

$$\alpha = 1 - |R|^2 \tag{20}$$

3.7. Determination of the Loss Factor

To calculate the propagation constant and characteristic impedance of the powder layer, as described in Section 3.2, determining the loss factor η is necessary. As the loss factor changes, the SAC at the primary peak frequency also changes. Therefore, the loss factor with the best agreement between the experimental and theoretical values is searched for when the loss factor varies from 0 to 1. Figure 10 presents the sensitivity analysis of the loss factor for granulated silica. The first-order peak of the sound absorption coefficient initially increases with the loss factor but gradually decreases beyond a certain point. A loss factor of 0.257 was selected for granulated silica, as it yielded the best agreement between theoretical and experimental sound absorption curves. The theoretical SAC values were calculated according to Section 3.6. Figure 11 shows the relationship between the loss factor and first-order peak SAC for each of the powders listed in Table 1. For granulated silica, hollow plastic beads A, hollow glass beads, and calcium-carbonate-coated hollow plastic beads A, B and C, two loss factors were obtained, for which the experimental and theoretical SACs matched. For hollow plastic beads B, only one loss factor was obtained. Table 2 shows the loss factors obtained for each powder, listed from the lower to the higher η_1 and η_2. The respective loss factors are verified in Section 4.

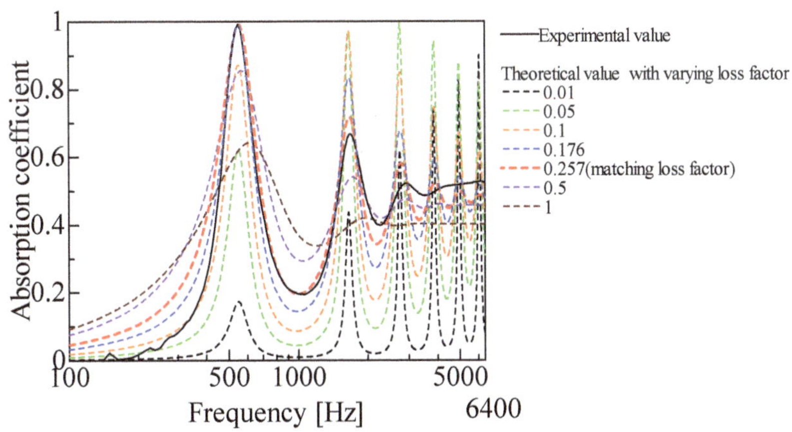

Figure 10. Sensitivity analysis of loss factor. (granulated silica).

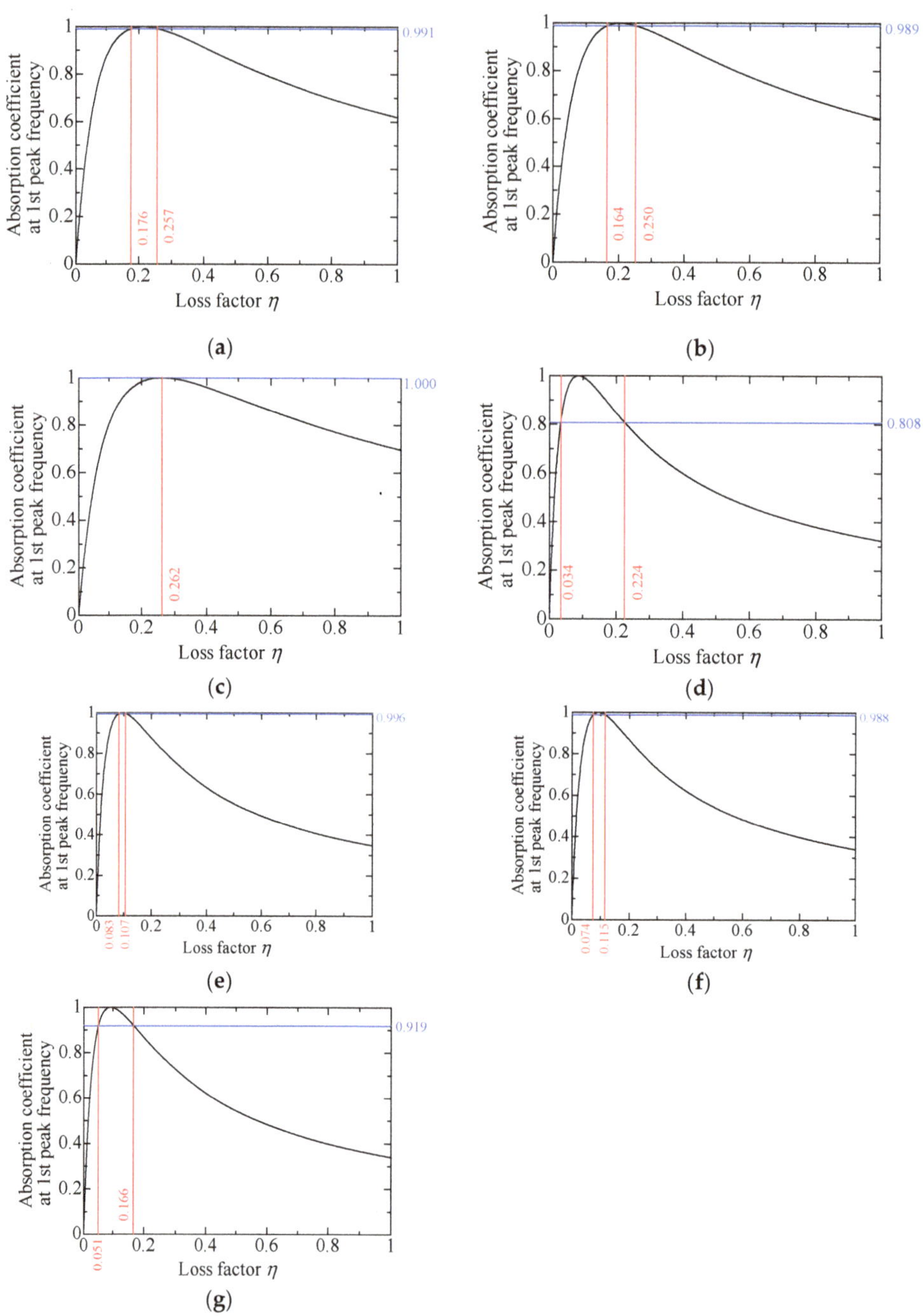

Figure 11. Estimation of loss coefficients consistent with experimental values at the first absorption peak: (**a**) granulated silica; (**b**) hollow plastic beads A (Expancel® 920DE40d30); (**c**) hollow plastic beads B (Matsumoto micro sphere® F–80DE); (**d**) hollow glass beads; (**e**) calcium-carbonate-coated hollow plastic beads A (EMC–40(B)); (**f**) calcium-carbonate-coated hollow plastic beads B (EMC–80(B)); (**g**) calcium-carbonate-coated hollow plastic beads C (EMC–120(α)).

Table 2. Peak SACs and loss factors.

	Experimental Value of the Absorption Coefficient at First Peak Frequency	η_1	η_2
Granulated silica	0.991	0.176	0.257
Hollow plastic beads A (Expancel® 920DE40d30)	0.989	0.164	0.250
Hollow plastic beads B (Matsumoto micro sphere® F-80DE)	1.000	0.262	
Hollow glass beads	0.808	0.034	0.224
Calcium-carbonate-coated hollow plastic beads A (EMC–40(B))	0.996	0.083	0.107
Calcium-carbonate-coated hollow plastic beads B (EMC–80(B))	0.988	0.074	0.115
Calcium-carbonate-coated hollow plastic beads C (EMC–120(α))	0.919	0.051	0.166

4. Verification of the Transfer Matrix of the Powder Layer

4.1. Comparison of the Experimental and Theoretical Values of the SAC in the Powder Layer Alone

In the transfer matrix of the powder obtained in Section 3, the validity of the transfer matrix was verified by multiplying the transfer matrix with other acoustic elements and comparing the experimental and theoretical values. Figure 12 shows the experimental and theoretical values of the SAC of the powder layer alone, as shown in Figure 4a. The powder layer thickness was 20 mm. In Figure 12a, the maximum, minimum, and mean values from 10 experimental trials with granulated silica are shown to validate the selection of the theoretical values. Near the first-order absorption peak, the theoretical curves align closely with the experimental upper and lower bounds, indicating that the model achieves a high level of accuracy. Theoretical values using the respective loss factors are given for granulated silica, hollow plastic beads A, hollow glass beads, and calcium-carbonate-coated hollow plastic beads A, B and C for which two loss factors were obtained. Figure 12a,b,d–g shows that the loss factors that tend to be closer to the experimental values are all η_2, based on the approximate shape of the sound absorption curves. In Figure 12c, only η_1 was determined. In both comparisons, the theoretical values for the powder layer alone reproduce the experimental values well if the appropriate loss factor is chosen. For this reason, these loss factors were used in the subsequent theoretical analysis.

Table 3 shows the percentage error of the first peak sound absorption frequency of the theoretical values compared to the experimental values. For all powders, the error was of less than 1%. This finding suggests that this study's transfer matrix can enable estimation of the powder layer SAC.

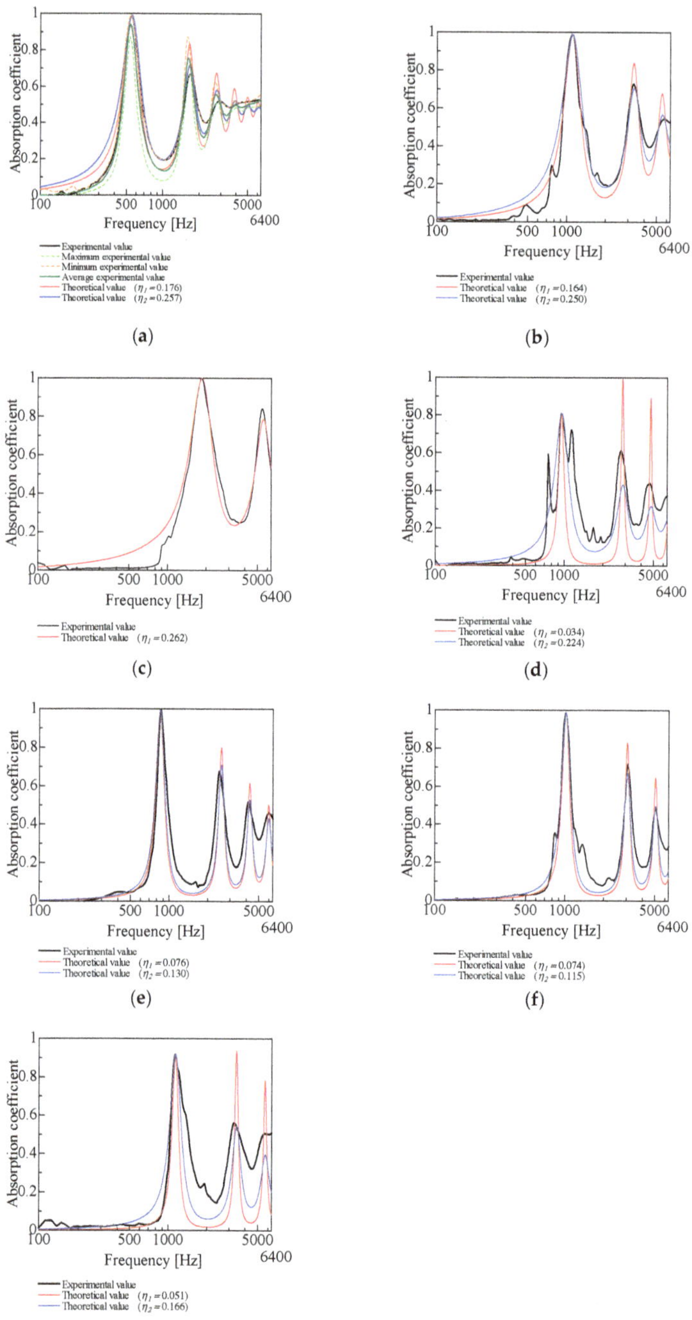

Figure 12. Comparison of experimental and theoretical values (l_p = 20 mm): (**a**) granulated silica; (**b**) hollow plastic beads A (Expancel® 920DE40d30); (**c**) hollow plastic beads B (Matsumoto micro sphere® F–80DE); (**d**) hollow glass beads; (**e**) calcium-carbonate-coated hollow plastic beads A (EMC–40(B)); (**f**) calcium-carbonate-coated hollow plastic beads B (EMC–80(B)); (**g**) calcium-carbonate-coated hollow plastic beads C (EMC–120(α)).

Table 3. Comparison of the experimental and theoretical values of f_{p1} (powder layer only).

	f_{p1} (Experimental) [Hz]	f_{p1} (Theoretical) [Hz]	Percent Error [%]
Granulated silica	550	555	0.90
Hollow plastic beads A (Expancel® 920DE40d30)	1100	1107.5	0.68
Hollow plastic beads B (Matsumoto micro sphere® F–80DE)	1837.5	1852.5	0.81
Hollow glass beads	950	955	0.52
Calcium-carbonate-coated hollow plastic beads A (EMC–40(B))	850	850	0.00
Calcium-carbonate-coated hollow plastic beads B (EMC–80(B))	1012.5	1012.5	0.00
Calcium-carbonate-coated hollow plastic beads C (EMC–120(α))	1137.5	1142.5	0.44

4.2. Comparison of Experimental and Theoretical Values for the SAC with a Back Air Space

Verification was conducted to determine whether appropriate theoretical values could be obtained by multiplying the transfer matrices of other acoustic materials with those of the powder layer. Specifically, the case where a back air space is installed by adding a nonwoven fabric as a support layer behind the powder layer, as shown in Figure 4b, is verified. The powder layer was 20 mm, and the air layer behind was 20 mm.

Figure 13 compares the experimental and theoretical values of the SACs. When an air space is installed behind porous material, the sound absorption peak often shifts to lower frequencies due to the increased overall thickness of the sound-absorbing structure. The theoretical values also shifted to the low-frequency side, following the transition of the experimental values to the low-frequency side (see Figure 13). This shows that the theoretical values were successfully obtained by adding the other acoustic elements of the transfer matrix of the powder layer. The present method is applicable to other lightweight powders not examined in this paper, provided they satisfy the conditions required for longitudinal vibration. As the sound absorption mechanism remains the same, such powders fall within the applicable scope of this model [7].

Discrepancies between the theoretical values derived using the 1D beam model and the experimental data are observed at higher frequencies. These deviations are attributed primarily to two factors: first, the loss coefficient assumed in this study is frequency-independent; second, boundary layer viscosity, which becomes more pronounced at higher frequencies, is not considered in the present model. Future improvements may involve incorporating frequency-dependent loss coefficients and extending the model to include boundary layer viscosity effects to enhance predictive accuracy.

Table 4 shows the experimental and theoretical first-peak sound absorption frequencies and the percentage error of the theoretical values relative to the experimental values. For hollow plastic beads A and B and calcium-carbonate-coated hollow plastic beads A and B, the percentage error was of less than 3%, but it exceeded 8% for granulated silica, hollow glass beads, and calcium-carbonate-coated hollow plastic beads C.

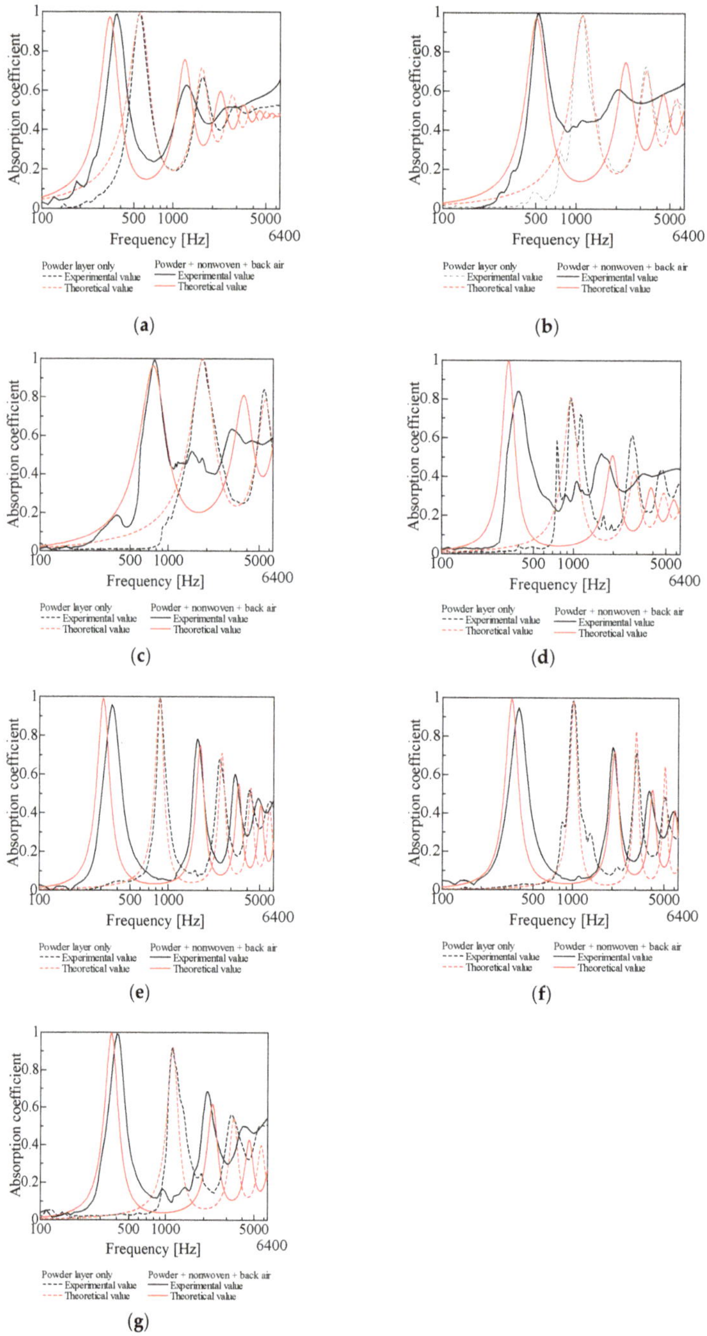

Figure 13. Comparison of experimental and theoretical values (l_p = 20 mm + nonwoven fabric + back air space 20 mm): (**a**) granulated silica; (**b**) hollow plastic beads A (Expancel® 920DE40d30); (**c**) hollow plastic beads B (Matsumoto micro sphere® F–80DE); (**d**) hollow glass beads; (**e**) calcium-carbonate-coated hollow plastic beads A (EMC–40(B)); (**f**) calcium-carbonate-coated hollow plastic beads B (EMC–80(B)); (**g**) calcium-carbonate-coated hollow plastic beads C (EMC–120(α)).

Table 4. Comparison of experimental and theoretical values of f_{p1} (powder layer + nonwoven + back air space).

	f_{p1} (Experimental) [Hz]	f_{p1} (Theoretical) [Hz]	Percent Error [%]
Granulated silica	375	330	12.0
Hollow plastic beads A (Expancel® 920DE40d30)	512.5	502.5	1.95
Hollow plastic beads B (Matsumoto micro sphere® F–80DE)	775	755	2.58
Hollow glass beads	387.5	320	17.4
Calcium-carbonate-coated hollow plastic beads A (EMC–40(B))	362.5	312.5	13.8
Calcium-carbonate-coated hollow plastic beads B (EMC–80(B))	387.5	340	12.3
Calcium-carbonate-coated hollow plastic beads C (EMC–120(α))	400	365	8.75

4.3. Powders to Which This Method Can Be Applied

The 1D beam model used in this study is intended for theoretical analysis of noise absorption owing to vibration, which is characteristic of lightweight, fine powders. Therefore, this method is not applicable to heavy powders with large grain sizes, as their sound absorption predominantly arises from boundary layer viscosity effects.

As an illustrative example, Figure 14 compares theoretical predictions from this model with experimental results for solid glass beads with a diameter of 0.8 mm and a bulk density of 1.55 g/cm^3. As shown in Figure 14, the model does not account for sound absorption due to boundary viscosity [5], which limits its applicability to larger and denser powders, such as solid glass beads.

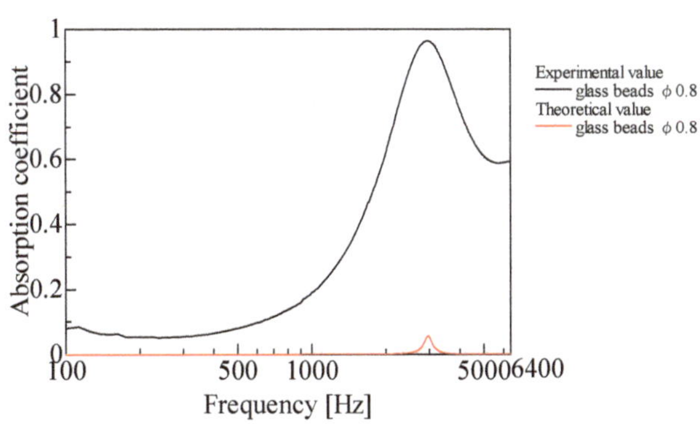

Figure 14. Comparison between theoretical and experimental results for solid glass beads (diameter: 0.8 mm; bulk density: 1.55 g/cm^3).

5. Conclusions

To estimate the SAC when the powder layers were laminated with various sound-absorbing structures, including a back-air layer, the powder layers were modeled as a continuum, and the transfer matrix was derived and validated. As a result, the following conclusions were reached:

(1) The powder layer was modeled as a 1D beam vibrating longitudinally. The transfer matrix was derived by analogy with an acoustic structure with the same governing equations.

(2) The theoretical values of the SAC of the powder layer alone were calculated using the transfer matrix obtained above and compared with the experimental values, which were close to each other.

(3) A comparison was made between the experimental and theoretical values for the case where the powder layer had a back-air layer. As a result, a shift in the peak sound absorption frequency to a lower frequency caused by adding a back air layer was confirmed in the theoretical and experimental values.

Author Contributions: Conceptualization, S.S.; Software, H.S. and Y.K. (Yosuke Kubo); Formal Analysis, Y.K. (Yuya Kawakami); Data Curation, Y.K. (Yuya Kawakami), H.S. and Y.K. (Yosuke Kubo); Supervision, S.S.; Project Administration, S.S. All authors have read and agreed to the published version of the manuscript.

Funding: This work was supported by JSPS KAKENHI, grant numbers 20K04359 and 24K07374.

Institutional Review Board Statement: Not applicable.

Data Availability Statement: The original contributions presented in this study are included in this article; further inquiries can be directed to the corresponding author.

Acknowledgments: The authors would like to thank Japan Fillite Co., Ltd., Matsumoto Yushi-Seiyaku Co., Ltd., for providing the powders for this study.

Conflicts of Interest: The authors declare no conflicts of interest.

Abbreviations

SAC	Sound absorption coefficient
SAM	Sound-absorbing materials
5-DOF	Five degrees of freedom

References

1. Okudaira, Y.; Kurihara, Y.; Ando, H.; Satoh, M.; Miyanami, K. Sound absorption measurements for evaluating dynamic physical properties of a powder bed. *Powder Technol.* **1993**, *77*, 39–48. [CrossRef]
2. Falegnami, A.; Tomassi, A.; Gunella, C.; Amalfitano, S.; Corbelli, G.; Armonaite, K.; Fornaro, C.; Giorgi, L.; Pollini, A.; Caforio, A.; et al. Defining conceptual artefacts to manage and design simplicities in complex adaptive systems. *Heliyon* **2024**, *10*, e41033. [CrossRef] [PubMed]
3. Okudaira, Y.; Kurihara, Y.; Ando, H.; Satoh, M.; Miyanami, K. Sound Absorption Characteristics of Powder Beds Comprised of Binary Powder Mixtures. *J. Jpn. Soc. Powder Powder Metall.* **1995**, *42*, 239–244. [CrossRef]
4. Sakamoto, S.; Yamaguchi, K.; Ii, K.; Takakura, R.; Nakamura, Y.; Suzuki, R. Theoretical and experiment analysis on the sound absorption characteristics of a layer of fine lightweight powder. *J. Acoust. Soc. Am.* **2019**, *146*, 2253–2262. [CrossRef] [PubMed]
5. Sakamoto, S.; Takakura, R.; Suzuki, R.; Katayama, I.; Saito, R.; Suzuki, K. Theoretical and experimental analyses of acoustic characteristics of fine-grain powder considering longitudinal vibration and boundary layer viscosity. *J. Acoust. Soc. Am.* **2021**, *149*, 1030–1040. [CrossRef] [PubMed]
6. Sakamoto, S.; Saito, R.; Jindai, K.; Ikeda, K. Conditions for sound absorption caused by longitudinal vibration of lightweight powder. *Noise Control Eng. J.* **2024**, *72*, 402–411. [CrossRef]
7. Sakamoto, S.; Jindai, K.; Ikeda, K.; Kawakami, Y.; Soeta, H. Predicting Lightweight Powders with Useful Sound Absorption Characteristics from Their Specifications. *Appl. Sci.* **2024**, *14*, 9765. [CrossRef]
8. Palacio, O.; Malfait, W.J.; Michel, S.; Barbezat, M.; Mazrouei-Sebdani, Z. Vibration and structure-borne sound isolation properties of silica aerogels. *Constr. Build. Mater.* **2023**, *399*, 132568. [CrossRef]
9. Biot, M.A. Theory of Propagation of Elastic Waves in a Fluid-Saturated Porous Solid. I. Low-Frequency Range. *J. Acoust. Soc. Am.* **1956**, *28*, 168–178. [CrossRef]

10. Tsuruha, T.; Yamada, Y.; Otani, M.; Takano, Y. Effect of casing on sound absorption characteristics of fine spherical granular material. *J. Acoust. Soc. Am.* **2020**, *147*, 3418–3428. [CrossRef] [PubMed]
11. Sakamoto, S.; Ikeda, K.; Kawakami, Y.; Soeta, H. Estimation of transfer matrix by applying the theoretical acoustic incident impedance of a lightweight powder layer to the two-thickness method. *Noise Control Eng. J.* **2025**, *73*, 9. [CrossRef]
12. Saiki, S.; Matsumura, T.; Fujise, A. Speaker System. European Patent EP 2003924A1, 25 April 2012.
13. *ISO 10534-2*; Acoustics—Determination of Acoustic Properties in Impedance Tubes—Part 2: Two-Microphone Technique for Normal Sound Absorption Coefficient and Normal Surface Impedance. International Organization for Standardization: Geneva, Switzerland, 2023.
14. Merkus, H.G. *Particle Size Measurements: Fundamentals, Practice, Quality*; Springer Science and Business Media: New York, NY, USA, 2019; p. 15.
15. Lahe, A.; Braunbrück, A.; Klauson, A. Modified transfer matrix method for steady-state forced vibration: A system of bar elements. *Proc. Est. Acad. Sci.* **2020**, *69*, 143–161. [CrossRef]
16. Ceaușu, V.; Craifaleanu, A.; Dragomirescu, C. Transfer matrix method for forced vibrations of bars. *UPB Sci. Bull. Ser. D Mech. Eng.* **2010**, *72*, 35–42.
17. Sugi, C. *Resonant Frequencies and Equivalent Mass Coefficients of Longitudinally Vibrating Composite Bars*; Faculty of Engineering, Yamanashi University: Yamanashi, Japan, 1964; Volume 14, pp. 101–109. (In Japanese) [CrossRef]
18. Laganà, F.; Prattico, D.; De Carlo, D.; Oliva, G.; Pullano, S.A.; Calcagno, S. Engineering Biomedical Problems to Detect Carcinomas: A Tomographic Impedance Approach. *Eng* **2024**, *5*, 1594–1614. [CrossRef]
19. Delany, M.E.; Bazley, E.N. Acoustic Properties of Fibrous Absorbent Materials. *Appl. Acoust.* **1970**, *3*, 105–116. [CrossRef]
20. Miki, Y. Acoustical properties of porous materials—Modifications of Delany-Bazley models. *J. Acoust. Soc. Jpn.* **1990**, *11*, 19–24. [CrossRef]
21. Komatsu, T. Improvement of the Delany-Bazley and Miki models for fibrous sound-absorbing materials. *Acoust. Sci. Technol.* **2008**, *29*, 121–129. [CrossRef]

materials

MDPI

Article

Digitally Controlled Piezoelectric Metamaterial for Low-Frequency and High-Efficiency Sound Absorption

Xiaodong Zhang [1,*], Jing Nie [1], Jinhong He [1], Fengbin Lin [1] and Yang Liu [2,*]

[1] College of Mechanical Engineering and Automation, Huaqiao University, Xiamen 361021, China; 23014080068@stu.hqu.edu.cn (J.N.); 23014080029@stu.hqu.edu.cn (J.H.); 24014080034@stu.hqu.edu.cn (F.L.)

[2] Jilin Jinheng Auto Parts Co., Ltd., Jilin 132000, China

* Correspondence: zhangxd@hqu.edu.cn (X.Z.); liuyang@jinhengqiche.cn (Y.L.)

Abstract: This study proposes a membrane-type metamaterial with digitally controlled piezoelectric actuation for low-frequency sound absorption applications. The hybrid structure integrates an aluminum membrane functionally bonded with programmable piezoelectric patches (PZTs) and a sealed air cavity. Two innovative control strategies—Resistance Enhancement and Resonance Enhancement—dynamically adjust circuit impedance to maximize electromechanical energy conversion efficiency, thereby optimizing absorption at targeted frequencies. These strategies are implemented via a real-time digital feedback system. A coupled piezoelectric-structural-acoustic model is established to characterize the system's transfer function, with validation through both finite element simulations and impedance tube experiments. Numerical and experimental results demonstrate nearly complete absorption around the resonant frequency, and the bandwidth can be further broadened through multi-resonance superposition. Theoretical analysis confirms that the active control strategies simultaneously modulate the acoustic impedance components (resistance and reactance), thereby optimizing electromechanical energy conversion efficiency. This work establishes a novel active-control methodology for low-frequency and high-efficiency noise mitigation.

Keywords: acoustic metamaterial; sound absorption; piezoelectric patch; resonant frequency; digital control

1. Introduction

There is an escalating prevalence of low-frequency noise pollution in industrial and urban environments, such as automotive engine compartments, air conditioning systems, subway stations, and industrial machinery rooms, where low-frequency noise (100–500 Hz) from rotating machinery, fluid dynamics, or structural vibrations poses significant health and environmental challenges. Traditional noise control methods utilize porous fibers, foams, and microperforated plates [1,2], rely on viscous-thermal dissipation mechanisms through submillimeter-scale pores, and exhibit frequency-dependent absorption coefficients that decline significantly in low frequency [3]. Therefore, manufacturing efficient, compact, low-frequency broadband absorbing materials poses a significant challenge.

Acoustic metamaterials, defined as artificial structures with subwavelength-scale geometries, enable unconventional sound manipulation beyond the capabilities of natural materials, such as a negative refractive index or subwavelength absorption [3–5]. As structural materials, acoustic metamaterials can manipulate sound by adjusting structures at deep subwavelength scales. Thus, well-engineered metamaterials can create low-frequency bandgaps without bulky modules for low-frequency sound isolation or reduction. For instance, spatial spiral design can produce considerable sound absorption effects at subwavelength thickness [6–9], while the metamaterials formed by the Helmholtz structures also confer commendable sound absorption performance at their resonance [10–13]. By varying the geometry and assembling metamaterials at different operating frequencies, this strategy could widen their sound absorption frequency range [14–16]. Moreover, perfect sound reduction performance can also be achieved by designing a thin membrane on top of air cavities [17–19]. The absorption frequency can be reduced by adhering small mass blocks to the membrane, resulting in a more compact structure with thicknesses much smaller than the operating wavelength. These research findings indicate promising potential for vibration and noise control, yet modifications in frequency range post-structural fixation present limitations as the operating frequency is inherently tied to physical geometry.

Multifield-coupled active metamaterials offer dynamic tunability through external field modulation [20,21]. Among these, piezoelectric materials stand out for their small size, light weight, and compatibility with electronic control circuits. Pioneering studies have implemented shunted piezoelectric networks for noise and vibration control [22–25]. Unlike passive acoustic materials (e.g., porous foams), which rely on inherent visco-thermal losses, piezoelectric metamaterials offer active tuning through piezoelectric coupling, enabling dynamic adjustment of absorption frequency without geometric modifications. Passive absorbers with piezoelectric membranes or patches can facilitate single or multi-modal sound insulation capabilities at tuned resonant frequencies controlled by different circuits, and their sound insulation frequency range can be varied by adjusting shunt circuit parameters [26–28]. Notably, Liao's work [29] proposes an adaptive absorption method incorporating negative capacitors and inductors. However, its practical implementation is constrained by technical challenges in constructing simultaneous negative capacitor–inductor circuits, which often induce system instability, high costs, and complex debugging burdens [30,31]. To address these challenges, Wang et al. [32,33] introduced two active enhancement strategies—the resonance-amplifier and amplifier-resonance strategies—for improved low-frequency vibration attenuation. Subsequently, Wang et al. [34] implemented these active control mechanisms in a metamaterial beam using digital control technology, achieving multi-modal vibration suppression. Meanwhile, recent advancements in the digital control of piezoelectric shunts have expanded metamaterials' tunability [35–39]. For instance, Yan et al. [40] provided a comprehensive review of shunt damping technologies, highlighting the importance of impedance modulation in achieving broadband noise suppression. Marakakis et al. [41] systematically compared various piezoelectric shunting strategies, demonstrating the superiority of active enhancement methods in vibration control. Additionally, Zheng et al. [42] conducted both analytical and experimental investigations on a piezoelectric meta-shell shunted with digital circuits, achieving programmable multiple bandgaps and effectively suppressing structural vibration and sound radiation.

In summary, acoustic metamaterials shunted with piezoelectric materials demonstrate significant potential for efficient and adjustable sound absorption. However, related research remains insufficient. Inspired by these advancements, this paper introduces a membrane-type metamaterial shunted with digitally controlled piezoelectric patches (PZTs). By employing a novel digital control strategy, the proposed metamaterial achieves adjustable, near-perfect sound absorption at programmable frequencies, offering a promising solution for broadband noise control.

The subsequent sections are organized as follows: Section 2 details the design and theoretical model of the proposed metamaterial, including its structural configuration and governing equations. Section 3 presents an acoustic coupling analysis model based on equivalent medium theory and finite element methods, elucidating sound absorption mechanisms through the relationship between resistance and acoustic factors. Section 4 employs numerical analysis and experimental validation to investigate sound absorption performance, providing insights into the underlying physical mechanisms. Finally, Section 5 summarizes the key findings from both theoretical and experimental results, highlighting the practical implications and future research directions of this work.

2. Physical Model and Theory

2.1. Design of the Metamaterial

This section first introduces a hybrid structure combining an aluminum membrane and piezoelectric patches (PZTs), termed a 'membrane-type metamaterial' here. This design leverages the vibration of a membrane and PZTs, as well as air cavity resonance to achieve low-frequency sound absorption. The design principles and active control circuits are then elaborated. Subsequently, effective medium theory is employed to derive the effective bending parameters of the membrane shunted with the circuit, establishing analytical models for piezoelectric-structural-acoustic interactions that incorporate the active enhancement control mechanism. Ultimately, a finite element analysis model is constructed based on these equivalent parameters, enabling detailed numerical simulations of the metamaterial's vibration and sound absorption performance.

Figure 1 presents the conceptual design of the proposed acoustic metamaterial and its corresponding digital control circuit. The metamaterial consists of two key components: an aluminum membrane with PZTs bonded on each side, serving as a sensing unit and actuating unit, respectively; and a cavity with a depth of 20 mm behind the composite membrane. The PZTs are polarized in opposite directions along the z-axis, interfacing with a digital controller for active piezoelectric shunt control. The control circuit comprises a charge amplifier, an analog-to-digital converter (ADC), a microprocessor, a digital-to-analog converter (DAC), and a proportional amplifier. Mechanical stress from PZT a is converted into a voltage signal V_1 through the charge amplifier, and further digitized by the ADC. Using these digital signals, the microprocessor calculates the output signal V_2 in real-time based on a specific transfer function. The DAC then converts the digital signal into an analog voltage signal V_2, which is amplified by a factor of β and applied to PZT b via the proportional amplifier, achieving active piezoelectric shunt control according to the designed transfer function. The geometric and material parameters of the proposed metamaterial are listed in Tables 1 and 2.

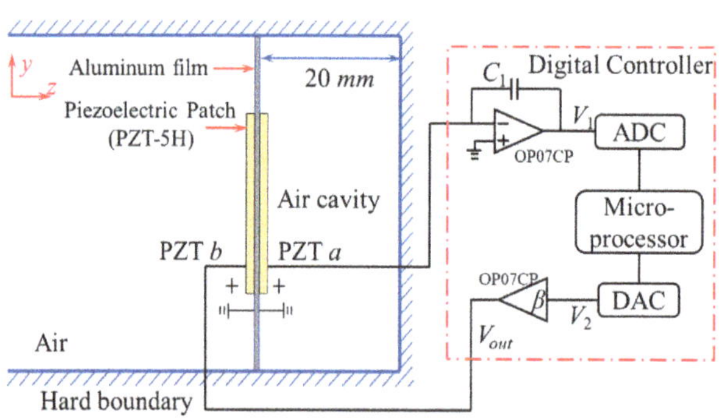

Figure 1. Sketch of the proposed membrane-type metamaterial shunted with a digital circuit.

Table 1. Geometric and physical parameters of membrane.

Parameters	Value
Length	$l_{Al} = 80$ mm
Thickness	$h_{Al} = 0.17$ mm
Density	$\rho_{Al} = 2700$ kg·m^{-3}
Young's modulus	$Y_{Al} = 69.5$ GPa
Poisson's ratio	$v_{Al} = 0.3$

Table 2. Geometric and physical parameters of PZT.

Parameters	Value
Length	$l_p = 40$ mm
Thickness	$h_p = 0.25$ mm
Density	$\rho_p = 7600$ kg·m^{-3}
Compliance coefficient	$S_{11}^E = 1.85 \times 10^{-12}$ m^2·N^{-1} $S_{12}^E = 4.72 \times 10^{-12}$ m^2·N^{-1}
Piezoelectric strain constant	$d_{31} = -101 \times 10^{-12}$ C·N^{-1}
Dielectric constant	$\varepsilon_{33}^T = 1.8 \times 10^{-8}$ F·m^{-1}

2.2. Design of the Transfer Function

During acoustic simulation, the PZTs bonded to the membrane undergo deformation, generating strain that drives the piezoelectric effect. This necessitates the inclusion of strain effects in our analysis. Accordingly, the membrane region bonded with the PZT is modeled as a thin flexible plate, and its equivalent elastic modulus is calculated using classical laminated plate theory to derive the transfer function required by the digital controller.

As shown in Figure 1, we define the z-axis as perpendicular to the electrode surface on the PZT (parallel to the aluminum membrane). The constitutive equation of the PZT is [26] as follows:

$$\begin{cases} S_1 = s_{11}^E T_1 + s_{12}^E T_2 + d_{31} E_3, \\ S_2 = s_{12}^E T_1 + s_{11}^E T_2 + d_{31} E_3, \\ S_6 = 2(s_{11}^E - s_{12}^E) T_6, \\ D_3 = d_{31} T_1 + d_{31} T_2 + \varepsilon_{33}^T E_3, \end{cases} \quad (1)$$

where S_1, T_1, S_2, and T_2 denote normal strains and stresses along x and y axes; S_6 and T_6 represent shear strains and stresses perpendicular to the x axis and along the y axis, respectively. D_3 denotes the electric displacement on the electrode; E_3 represents the electric field intensity; S_{11}^E and S_{12}^E symbolize flexibility coefficients of a piezoelectric material under constant field strength; ε_{33}^T represents the permittivity of the piezoelectric material under static strain; and d_{31} is the piezoelectric constant.

The piezoelectric coupling effect, described by Equation (1), establishes a bidirectional energy conversion between mechanical deformation of the membrane and electrical signals in the PZTs. This mechanism is distinct from electromagnetic coupling as it relies on electrostatic interactions within the piezoelectric material rather than magnetic fields, making it suitable for low-frequency acoustic applications.

PZT a, serving as a sensor unit, equates ground, hence its field intensity E_3 is zero, which simplifies its piezoelectric equation. The constitutive equation of PZT a can be expressed as follows:

$$\begin{Bmatrix} S_1^a \\ S_2^a \\ S_6^a \end{Bmatrix} = \begin{bmatrix} S_{11}^E & S_{12}^E & 0 \\ S_{12}^E & S_{11}^E & 0 \\ 0 & 0 & 2(S_{11}^E - S_{12}^E) \end{bmatrix} \begin{Bmatrix} T_1 \\ T_2 \\ T_6 \end{Bmatrix} \tag{2}$$

Obviously, Equation (2) resembles the standard stress–strain formula for isotropic plates:

$$\begin{Bmatrix} S_1 \\ S_2 \\ S_6 \end{Bmatrix} = \begin{bmatrix} \frac{1}{E_p} & -\frac{v_p}{E_p} & 0 \\ -\frac{v_p}{E_p} & \frac{1}{E_p} & 0 \\ 0 & 0 & \frac{1}{G_p} \end{bmatrix} \begin{Bmatrix} T_1 \\ T_2 \\ T_6 \end{Bmatrix}; \tag{3}$$

Hence, the equivalent parameter of PZT a can be expressed as follows:

$$\begin{cases} E_p^a = \dfrac{1}{S_{11}^E} \\ v_p^a = -\dfrac{S_{12}^E}{S_{11}^E} \end{cases} \tag{4}$$

Hence, the equivalent parameter of PZT a can be expressed as follows:

$$I_a = -A_s D_s^a s \tag{5}$$

where $s = i\omega$ denotes the Laplacian operator; and A_s is the area of the PZT's electrodes. The correlation between voltage V_1 and current I_a is expressed as follows:

$$V_1 = -I_a Z_1 = A_s D_s^a Z_1 s \tag{6}$$

The capacitance's impedance value, Z_1, in the charge amplifier is given by $Z_1 = 1/(C_1 s)$. By solving Equations (2) and (6), the relationship between the internal mechanical strain S_1^a, S_2^a and voltage V_1 is as follows:

$$V_1 = \frac{A_s d_{31} Z_1 s}{S_{11}^E S_{12}^E}(S_1^a + S_2^a) = G(S_1^a + S_2^a) \tag{7}$$

Herein, $G = A_s d_{31} Z_1 s / S_{11}^E S_{12}^E$. From the above equation, it is evident that the voltage V_1 is directly proportional to strains S_1^a and S_2^a with a constant ratio. Hence, V_1 serves as an ideal sensor for strain in PZT a. At the subwavelength scale, the strain across PZT a and b can be approximated as uniform and reciprocally constrained:

$$S_1^a = -S_1^b;\ S_2^a = -S_2^b \tag{8}$$

The output voltage V_{out} and current I_b of PZT b satisfy the following relationship:

$$\begin{cases} V_{out} = E_3^b h_p \\ I_b = A_s D_3^b s = \dfrac{\alpha V_1 - V_{out}}{Z_2} \end{cases} \tag{9}$$

where α denotes C_p/C_1, and Z_2 signifies the impedance of the circuit implemented by the microcontroller. Employing Equation (2), the constitutive equation for PZT b can be reformulated as follows:

$$\begin{cases} S_1^b + S_2^b = (T_1^b + T_2^b)(S_1^b + S_2^b) + 2d_{31}E_3^b \\ D_3^b = d_{31}(T_1^b + T_2^b) + \varepsilon_{33}^T E_3^b \end{cases} \tag{10}$$

Subtracting stresses T_1^a and T_2^a from the above equation yields the following:

$$\frac{S_1^b + S_2^b}{S_{11}^E + S_{12}^E} = \frac{D_3^b}{d_{31}} - \frac{\varepsilon_{33}^T}{d_{31}}E_3^b + \frac{2d_{31}}{S_{11}^E + S_{12}^E}E_3^b \tag{11}$$

Merging Equations (7), (9), and (11) yields the relationship between V_1 and V_{out}:

$$V_1\left(\frac{Z_2}{Z_1} + \alpha\right) = \frac{V_{out}(h_p - A_s d_{31} Z_2 s)}{h_p}\left(\frac{2d_{31}}{S_{11}^E + S_{12}^E} - \frac{\varepsilon_{33}^T}{d_{31}}\right) \tag{12}$$

Next, a parameter is defined:

$$F = \frac{A_s d_{31} Z_2 s}{h_p}\left(\frac{2d_{31}}{S_{11}^E + S_{12}^E} - \frac{\varepsilon_{33}^T}{d_{31}}\right) \tag{13}$$

Then, Equation (12) can be expressed as follows:

$$\frac{V_{out}}{V_1} = \frac{Z_2/Z_1 + \alpha}{1 - FZ_2 s} = H(s) \tag{14}$$

Combining Equations (2), (8) and (11), the constitutive equation of PZT b is expressed as follows:

$$\begin{cases} S_1^b = S_{11}^E T_1 + S_{12}^E T_2 - \dfrac{d_{31}GH(s)}{h_p}(S_1^b + S_2^b) \\ S_2^b = S_{12}^E T_1 + S_{11}^E T_2 - \dfrac{d_{31}GH(s)}{h_p}(S_1^b + S_2^b) \\ S_6^b = 2(S_{11}^E - S_{12}^E)T_6 \end{cases} \tag{15}$$

Defining $A = d_{31}GH(s)/h_p$, the equation can be simplified as follows:

$$\begin{cases} S_1^b = T_1\dfrac{S_{11}^E + A(S_{11}^E - S_{12}^E)}{1+2A} + T_2\dfrac{S_{12}^E + A(S_{12}^E - S_{11}^E)}{1+2A} \\ S_2^b = T_1\dfrac{S_{12}^E + A(S_{12}^E - S_{11}^E)}{1+2A} + T_2\dfrac{S_{11}^E + A(S_{11}^E - S_{12}^E)}{1+2A} \\ S_6^b = 2(S_{11}^E - S_{12}^E)T_6 \end{cases} \tag{16}$$

Similarly, by comparing with Equation (3), the equivalent elastic parameters of PZT b are derived as follows:

$$\begin{cases} E_p^b = \dfrac{1+2A}{S_{11}^E + A(S_{11}^E - S_{12}^E)} \\ v_p^b = -\dfrac{S_{12}^E + A(S_{12}^E - S_{11}^E)}{S_{11}^E + A(S_{11}^E - S_{12}^E)} \end{cases} \tag{17}$$

Therefore, PZT b can be modeled as a homogeneous isotropic elastic medium, exhibiting frequency-dependent material properties, which can be computed using the aforementioned equations.

As illustrated in Figure 1, the amplification ratio of the amplifier circuit is β, governing the relationship between the output voltage V_{out} and the controlled voltage V_2, which satisfies the following:

$$V_{out} = \beta V_2 \tag{18}$$

The complete analog transfer function of the digital controller can be expressed as $H_{DP}(s) = V_2/V_1$. By connecting Equations (14) and (18), we derive the following governing equation for the control system:

$$H_{DP}(s) = \frac{H(s)}{\beta} = \frac{Z_2/Z_1 + \alpha}{\beta(1 - FZ_2 s)} \tag{19}$$

3. Finite Element Modeling and Experimental Setup

To evaluate the acoustic absorption efficiency of the proposed metamaterial, a multi-physics coupling model was developed using COMSOL Multiphysics v6.1, integrating piezoelectricity, circuits, elastic structures, and acoustics. Figure 2 illustrates the finite element model for an incident wave frequency of 200 Hz and amplitude of 1 Pa, along with the corresponding sound pressure distribution. The simulation is driven by a background pressure field generating a plane wave propagating to the right. A perfect matching layer (PML) is positioned to the left of the pressure field to absorb reflected waves, while the remaining boundaries are modeled as hard sound field boundaries to simulate a rigid-walled waveguide, which is consistent with the acoustic impedance tube environment.

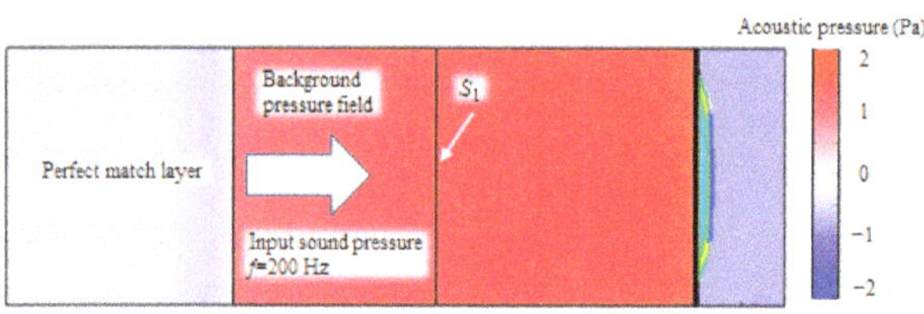

Figure 2. Finite element model and sound pressure distribution in tube.

To determine the sound absorption coefficient of the proposed metamaterial, we calculated the incident sound power (W_{in}) and reflected sound power (W_{ref}) on plane S_1:

$$W_{in} = \int_{S_1} \frac{{P_{in}}^2}{2\rho_0 c_0} dS \tag{20}$$

$$W_{ref} = \int_{S_1} \frac{{P_{ref}}^2}{2\rho_0 c_0} dS \tag{21}$$

The reflection coefficient is as follows:

$$R = \frac{W_{ref}}{W_{in}} \tag{22}$$

Thus, the sound absorption coefficient can be obtained as follows:

$$\alpha = 1 - |R|^2 \tag{23}$$

Meanwhile, the acoustic impedance Z_n of the metamaterial can be obtained by the following formula:

$$Z_n = \frac{p}{\mathbf{v} \cdot \mathbf{n}} = R_t + X_t i \tag{24}$$

Here, **v** denotes the sound velocity near the metamaterial, and **n** is the unit vector in the normal direction. The real part of the acoustic impedance represents the acoustic resistance R_t, while the imaginary part corresponds to the acoustic reactance X_t. The sound absorption coefficient α of the metamaterial is directly related to its acoustic impedance. Consequently, the sound absorption coefficient can be expressed in terms of R_t and X_t:

$$\alpha = \frac{4R_t}{(1+R_t)^2 + {X_t}^2} \tag{25}$$

Based on Equation (25), the relationship between sound absorption coefficient α, acoustic resistance R_t, and acoustic reactance X_t can be obtained. Figure 3 presents the isocline graph of α against R_t and X_t, illustrating that achieving a high sound absorption coefficient requires satisfying two conditions: $R_t \approx 1$ and $X_t \approx 0$. Hence, to optimize sound absorption in the proposed metamaterial, the circuit parameters must be tuned to ensure R_t and X_t approach these target values as closely as possible.

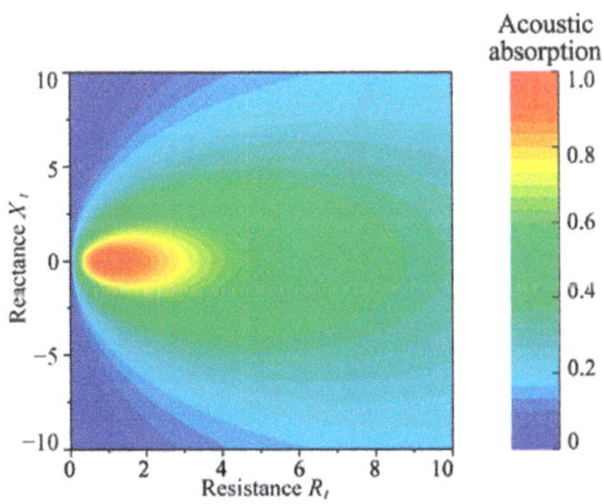

Figure 3. Graphs of the acoustic absorption coefficient α at varying values of R_t and X_t.

Figure 4 illustrates the experimental setup, with the inset subgraph detailing the configuration of the composite sample. The sample features an aluminum membrane concentrically sandwiched between two PZT-5H piezoelectric patches, which were bonded using epoxy adhesive (Loctite 330) with a curing process at room temperature (about 25 °C) for 12 h. The geometric dimensions and material properties of both the aluminum membrane and PZT elements, as specified in Tables 1 and 2, respectively, were provided by certified manufacturers: Hefei Xingfan Metal Materials Co., Ltd. (Hefei, China; aluminum components) and Baoding Hongsheng Acoustic Electronic Equipment Co., Ltd. (Baoding, China; piezoelectric elements). The positive and negative electrodes of the PZTs are connected to an external digital control circuit via copper foil outside the fixed frame. The low-cost digital control circuit integrates a Texas Instruments OPA454 precision operational amplifier (operational range: ±50 V) and an STM32F446 microprocessor equipped with two 12-bit DACs and multiple ADCs. The microprocessor was selected for its compatibility with low-frequency piezoelectric current control.

Sound absorption performance was evaluated using a custom-built acoustic impedance tube compliant with GB/T 18696.1-2004 (see [28] for design details). The tube, with a width matching the membrane dimension l_{Al} = 0.08 m, incorporates two BSWA TECH MPA416-2 microphones spaced 350 mm apart on the specimen's right side. A computer-generated white noise signal (100–500 Hz) is amplified by a power amplifier and used to drive a speaker at the right end of the tube, generating low-frequency noise. The signals captured by the microphones are converted into voltage signals by a signal conditioner (SKC Q882) and transmitted to a computer via a data acquisition card (National Instruments USB-4431). The sound absorption characteristics of the specimen are then calculated from the experimental data using custom-developed software.

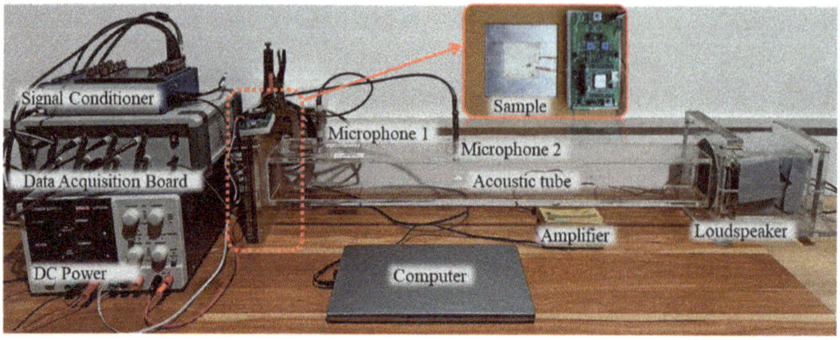

Figure 4. Photos of the experimental setup and the experimental sample.

4. Results and Discussion

4.1. "Resistance-Enhancement" Control

The acoustic absorption performance of the proposed metamaterial is investigated. The digital control strategy enhances circuit efficiency, improving energy conversion between mechanical and electrical domains, thereby increasing the electromechanical coupling coefficient. Initially, the control strategy is configured to implement "resistance-enhancement" functionality as illustrated in Figure 5. Resistance in the circuit facilitates sound wave energy dissipation, and the integration of active enhancement control further amplifies this dissipation capability. As an example, the amplification factor is set to $\beta = 10$ and the resistance to $R = 150\ \Omega$.

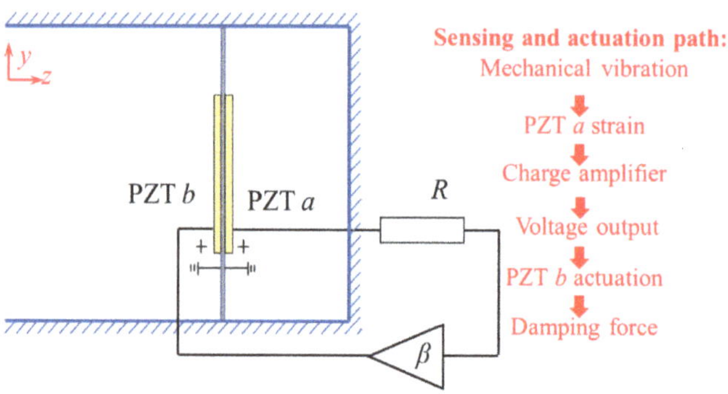

Figure 5. The resistance-enhancement control strategy.

Figure 6 shows the absorption curve of the metamaterial under the "resistance-enhancement" control strategy. For comparison, the absorption curve of the metamaterial shunted with a passive resistor is also provided. The thick dashed and solid lines represent the simulation and experimental results for the actively controlled metamaterial, while the thin dashed and solid lines correspond to the passively shunted metamaterial. As shown in Figure 6, the absorption coefficient is nearly zero for the passively shunted metamaterial, indicating its ineffectiveness in absorbing and dissipating sound wave energy. In contrast, under the "resistance-enhancement" control strategy, the metamaterial achieves an absorption coefficient of 0.97 at approximately 224 Hz, demonstrating near-total absorption. It is worth noting that the resonance exploited in this strategy originates from the mechanical vibration modes of the piezoelectric membrane (natural frequencies: 224 Hz), not from electrical resonance in the shunt circuit. The digital controller dynamically adjusts the equivalent resistance to match the mechanical impedance of the membrane at these resonant frequencies, thereby maximizing energy dissipation. The close agreement between experimental and simulated results validates the effectiveness of the proposed control strategy.

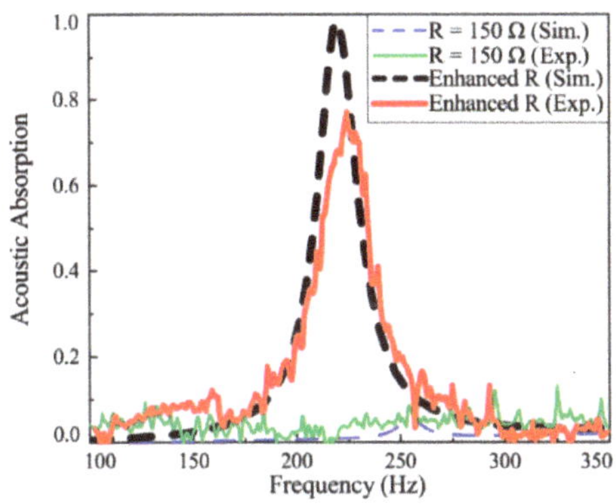

Figure 6. Sound absorption characteristics of the metamaterial with resistance-enhancement control strategy.

The digital controller modulates the transfer function $H_{DP}(s) = V_2/V_1$ according to Equations (15)–(17) to adjust the effective stiffness of PZT patches in real time. By adjusting parameters like the amplification factor β and circuit impedance Z_2, the acoustic resistance is optimized; $R_t \approx 1$ and $X_t \approx 0$.

The above discussion establishes the correlation between acoustic absorption and the acoustic resistance R_t and reactance X_t. Therefore, Equation (25) can be used to explain the exceptional acoustic absorption of the proposed metamaterial. Figure 7 illustrates the frequency-dependent variations in R_t and X_t for the "resistance-enhancement"-controlled metamaterial (thick line) and the passively shunted metamaterial (thin line).

As demonstrated in Figure 7a, the frequency response of R_t for the passively shunted metamaterial exhibits a sharp peak near $R_t = 1$, accompanied by X_t approaching infinity. This configuration violates the high absorption condition ($R_t \approx 1$, $X_t \approx 0$), resulting in suboptimal sound absorption performance. In contrast, under the "resistance-enhancement" control strategy, the R_t peak broadens significantly across the frequency range, and the associated X_t curve becomes smoother. Near $R = 1$, X_t approximates near zero, satisfying the high absorption requirement. Consequently, the actively controlled metamaterial achieves superior sound absorption performance.

It is noted that the frequency of near-perfect absorption coincides with the intrinsic resonance mode of the metamaterial. Near this resonance, the metamaterial undergoes significant deformation, amplifying the electromechanical coupling effect. The proposed control strategy enhances this coupling, improving energy conversion between mechanical and electrical domains, thereby increasing the effective resistance for sound wave dissipation. This mechanism enables the actively controlled metamaterial to achieve efficient sound absorption performance.

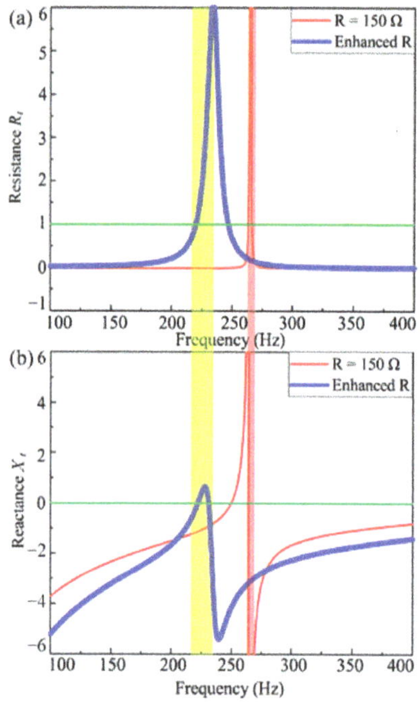

Figure 7. (**a**) The acoustic resistance and (**b**) the acoustic reactance curve of the metamaterial with resistance-enhancement control strategy.

4.2. "Resonance-Enhancement" Control

In "resistance-enhancement" control, to achieve a broader absorption bandwidth, additional resonances can be integrated into the metamaterial, enabling multiple resonance amplifications within the target frequency range. In this section, we implement the "resonance-enhancement" control strategy using the controller illustrated in Figure 8. The resistor R remained at 150 Ω, while the inductance L was set to 1.5 H, and the amplification factor β was fixed at 10.

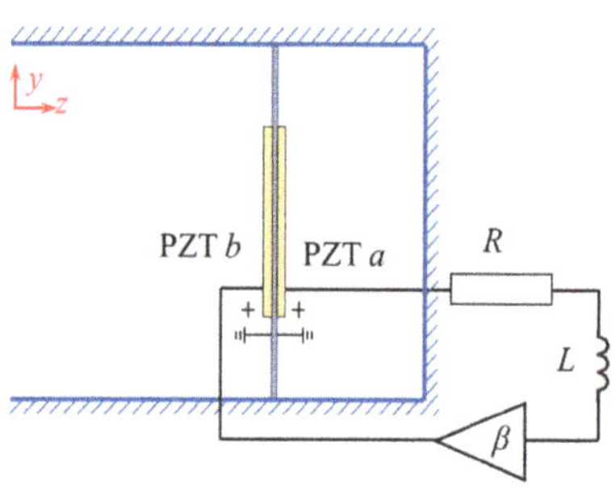

Figure 8. The resonance -enhancement control strategy.

Figure 9 illustrates the sound absorption properties of the "resonance-enhancement"-controlled metamaterial compared to the passively shunted metamaterial. Two prominent absorption peaks are observed near 170 Hz and 248 Hz, with absorption coefficients exceeding 0.96, indicating near-complete absorption at these frequencies. The simulation and experimental results show excellent agreement, validating the effectiveness of the proposed control strategy.

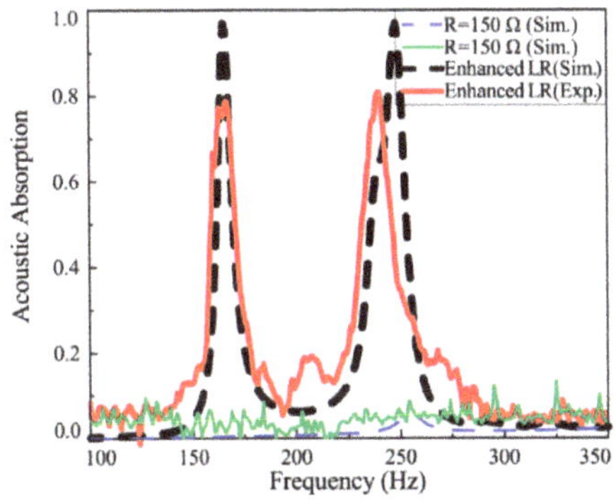

Figure 9. Sound absorption characteristics of the metamaterial with resonance-enhancement control strategy.

In Figure 10, under "resonance-enhancement" control, the acoustic resistance R_t exhibits two broad peaks in the frequency domain. One peak arises from the intrinsic resonance of the metamaterial, while the other is associated with the resonance generated by the control circuit. Notably, near R_t = 1 at these peaks, the corresponding acoustic reactance X_t becomes flat with values close to zero. This behavior indicates that the acoustic resistance and reactance at these frequencies satisfy the prerequisites for high sound absorption ($R_t \approx 1$, $X_t \approx 0$). Therefore, the metamaterial with the "resonance-enhancement" control strategy achieves a broader sound absorption bandwidth compared to other configurations, demonstrating its superior performance for broadband noise control.

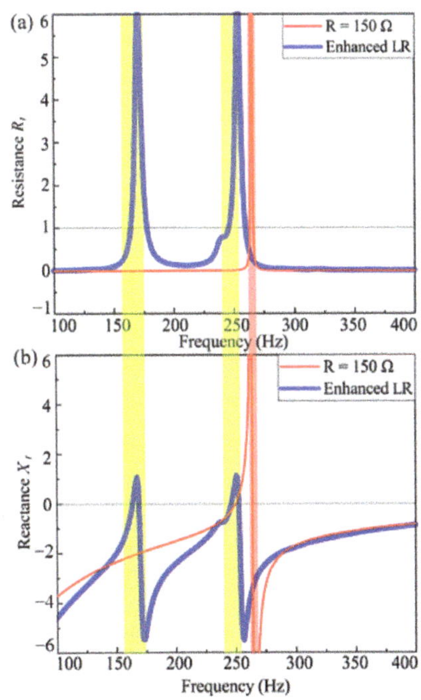

Figure 10. (**a**) The acoustic resistance and (**b**) the acoustic reactance curve of the metamaterial with resonance-enhancement control strategy.

Apparently, the proposed design exhibits superiority compared to the traditional design. Specially, passive designs (e.g., Helmholtz resonators [13,14] and membrane-based absorbers [18,19]) rely on fixed geometric resonances, making them ineffective for broadband or adaptive noise control. While passive metamaterials often require subwavelength thickness, our active design achieves comparable compactness (20 mm cavity depth) with a programmable frequency response.

5. Conclusions

This paper presents a digitally controlled membrane-type metamaterial for adjustable and broadband sound absorption. The proposed metamaterial consists of an aluminum membrane with two PZTs and an enclosed air cavity. The PZTs are shunted with a digital controlled circuit, enabling real-time tuning of the metamaterial's acoustic properties. A piezoelectric-structure-acoustic coupling model is derived to evaluate the sound absorption capability of the metamaterial. Through theoretical analysis and experimental validation, we demonstrate the exceptional absorption performance of the metamaterial under both "resistance-enhancement" and "resonance-enhancement" control strategies. Compared with passive acoustic metamaterials and passive piezoelectric shunts, the proposed digital control strategy offers superior stability, real-time tunability, and broadband performance. The results indicate that the metamaterial achieves high sound absorption performance at subwavelength scales, particularly in the low-frequency range and near-resonance frequencies. The control strategy not only enhances electromechanical coupling efficiency but also enables real-time tuning of R_t and X_t, overcoming the fixed-frequency limitation of passive metamaterials and achieving broadband absorption. Furthermore, the adjustable transfer function of the digital control circuit enables real-time tuning of sound absorption performance without modifying metamaterial geometry, facilitating a broader absorption bandwidth. The compact design and real-time tunability of the proposed metamaterial make it particularly suitable for space-constrained environments requiring broadband noise control, such as aerospace cabins, urban infrastructure, and precision manufacturing facilities.

Author Contributions: Conceptualization, X.Z. and Y.L.; methodology, J.H. and J.N.; software, J.H.; validation, F.L., J.N. and X.Z.; formal analysis, Y.L.; investigation, F.L.; resources, X.Z. and Y.L.; data curation, Y.L.; writing—original draft preparation, J.N.; writing—review and editing, Y.L. and X.Z.; visualization, Y.L.; supervision, X.Z.; project administration, X.Z.; funding acquisition, X.Z. All authors have read and agreed to the published version of the manuscript.

Funding: This work was supported by the National Natural Science Foundation of China (52305106), Fundamental Research Funds for the Central Universities (ZQN-1202), and Natural Science Foundation of Fujian Province (2022J05060).

Institutional Review Board Statement: Not applicable.

Informed Consent Statement: Not applicable.

Data Availability Statement: Data are contained within the article.

Acknowledgments: The authors thank the editors and anonymous reviewers for their valuable remarks and recommendations.

Conflicts of Interest: Author Yang Liu was employed by the company Jilin Jinheng Auto Parts Co., Ltd. The remaining authors declare that the research was conducted in the absence of any commercial or financial relationships that could be construed as a potential conflict of interest.

References

1. Tao, Y.; Ren, M.; Zhang, H.; Peijs, T. Recent Progress in Acoustic Materials and Noise Control Strategies—A Review. *Appl. Mater. Today* **2021**, *24*, 101141. [CrossRef]
2. Yang, M.; Sheng, P. Sound Absorption Structures: From Porous Media to Acoustic Metamaterials. *Annu. Rev. Mater. Res.* **2017**, *47*, 83–114. [CrossRef]
3. Zhang, X.; Qu, Z.; Wang, H. Engineering Acoustic Metamaterials for Sound Absorption: From Uniform to Gradient Structures. *iScience* **2020**, *23*, 101110. [CrossRef] [PubMed]
4. Ji, G.; Huber, J. Recent Progress in Acoustic Metamaterials and Active Piezoelectric Acoustic Metamaterials—A Review. *Appl. Mater. Today* **2022**, *26*, 101260. [CrossRef]
5. Qu, S.; Sheng, P. Microwave and Acoustic Absorption Metamaterials. *Phys. Rev. Appl.* **2022**, *17*, 047001. [CrossRef]
6. Cheng, Y.; Zhou, C.; Yuan, B.G.; Wu, D.J.; Wei, Q.; Liu, X.J. Ultra-Sparse Metasurface for High Reflection of Low-Frequency Sound Based on Artificial Mie Resonances. *Nat. Mater.* **2015**, *14*, 1013–1019. [CrossRef]
7. Li, Y.; Assouar, B.M. Acoustic Metasurface-Based Perfect Absorber with Deep Subwavelength Thickness. *Appl. Phys. Lett.* **2016**, *108*, 063502. [CrossRef]
8. Long, H.; Shao, C.; Cheng, Y.; Tao, J.; Liu, X. High Absorption Asymmetry Enabled by a Deep-Subwavelength Ventilated Sound Absorber. *Appl. Phys. Lett.* **2021**, *118*, 263502. [CrossRef]
9. Liu, X.; Duan, M.; Liu, M.; Xin, F.; Zhang, C. Acoustic Labyrinthine Porous Metamaterials for Subwavelength Low-Frequency Sound Absorption. *J. Appl. Phys.* **2021**, *129*, 195103. [CrossRef]
10. Lemkalli, B.; Fellah, Z.E.A.; Guenneau, S.; Lamzoud, K.; Mir, A.; Achaoui, Y. Bi-Functional Metamaterial Based on Helmholtz Resonators for Sound and Heat Insulation. *arXiv* **2023**, arXiv:2310.05098.
11. Ma, K.; Tan, T.; Yan, Z.; Liu, F.; Liao, W.-H.; Zhang, W. Metamaterial and Helmholtz Coupled Resonator for High-Density Acoustic Energy Harvesting. *Nano Energy* **2021**, *82*, 105693. [CrossRef]
12. Ramos, D.; Godinho, L.; Amado-Mendes, P.; Mareze, P. Experimental and Numerical Investigations of Ventilated Acoustic Metamaterial Based In-Parallel Arrangement of Helmholtz Resonator for Façade Screen. *INTER NOISE NOISE-CON Congr. Conf. Proc.* **2021**, *263*, 5382–5390. [CrossRef]
13. Romero-García, V.; Jiménez, N.; Theocharis, G.; Achilleos, V.; Merkel, A.; Richoux, O.; Tournat, V.; Groby, J.-P.; Pagneux, V. Design of Acoustic Metamaterials Made of Helmholtz Resonators for Perfect Absorption by Using the Complex Frequency Plane. *Comptes Rendus Phys.* **2020**, *21*, 713–749. [CrossRef]
14. Papadakis, N.M.; Stavroulakis, G.E. Tunable Helmholtz Resonators Using Multiple Necks. *Micromachines* **2023**, *14*, 1932. [CrossRef]
15. Chen, A.; Yang, Z.; Zhao, X.; Anderson, S.; Zhang, X. Composite Acoustic Metamaterial for Broadband Low-Frequency Acoustic Attenuation. *Phys. Rev. Appl.* **2023**, *20*, 014011. [CrossRef]
16. Jung, J.; Park, H.; Park, J.; Chang, T.; Shin, J. Broadband Metamaterials and Metasurfaces: A Review from the Perspectives of Materials and Devices. *Nanophotonics* **2020**, *9*, 3165–3196. [CrossRef]
17. Ma, G.; Sheng, P. Acoustic Metamaterials: From Local Resonances to Broad Horizons. *Sci. Adv.* **2016**, *2*, e1501595. [CrossRef]
18. Mei, J.; Ma, G.; Yang, M.; Yang, Z.; Wen, W.; Sheng, P. Dark Acoustic Metamaterials as Super Absorbers for Low-Frequency Sound. *Nat. Commun.* **2012**, *3*, 756. [CrossRef]
19. Huang, H.; Cao, E.; Zhao, M.; Alamri, S.; Li, B. Spider Web-Inspired Lightweight Membrane-Type Acoustic Metamaterials for Broadband Low-Frequency Sound Isolation. *Polymers* **2021**, *13*, 1146. [CrossRef]
20. Xu, J.; Yang, R.; Fan, Y.; Fu, Q.; Zhang, F. A Review of Tunable Electromagnetic Metamaterials With Anisotropic Liquid Crystals. *Front. Phys.* **2021**, *9*, 67. [CrossRef]
21. Wang, Y.-F.; Wang, Y.-Z.; Wu, B.; Chen, W.; Wang, Y.-S. Tunable and Active Phononic Crystals and Metamaterials. *Appl. Mech. Rev.* **2020**, *72*, 040801. [CrossRef]
22. Cheer, J.; Hook, K.; Daley, S. Active Feedforward Control of Flexural Waves in an Acoustic Black Hole Terminated Beam. *Smart Mater. Struct.* **2021**, *30*, 035003. [CrossRef]
23. Danawe, H.; Tol, S. Electro-Momentum Coupling Tailored in Piezoelectric Metamaterials with Resonant Shunts. *APL Mater.* **2023**, *11*, 091118. [CrossRef]
24. Jian, Y.; Hu, G.; Tang, L.; Shen, Y.; Zhan, Y.; Aw, K. Adaptive Piezoelectric Metamaterial Beam: Autonomous Attenuation Zone Adjustment in Complex Vibration Environments. *Smart Mater. Struct.* **2023**, *32*, 105023. [CrossRef]
25. Liao, Y.; Chen, Y.; Huang, G.; Zhou, X. Broadband Low-Frequency Sound Isolation by Lightweight Adaptive Metamaterials. *J. Appl. Phys.* **2017**, *123*, 091705. [CrossRef]
26. Zhang, H.; Wen, J.; Xiao, Y.; Wang, G.; Wen, X. Sound Transmission Loss of Metamaterial Thin Plates with Periodic Subwavelength Arrays of Shunted Piezoelectric Patches. *J. Sound Vib.* **2015**, *343*, 104–120. [CrossRef]
27. Zhang, H.; Xiao, Y.; Wen, J.; Yu, D.; Wen, X. Ultra-Thin Smart Acoustic Metasurface for Low-Frequency Sound Insulation. *Appl. Phys. Lett.* **2016**, *108*, 141902. [CrossRef]

28. Zhang, X.; Chen, F.; Chen, Z.; Wang, G. Membrane-Type Smart Metamaterials for Multi-Modal Sound Insulation. *J. Acoust. Soc. Am.* **2018**, *144*, 3514–3524. [CrossRef]
29. Liao, Y.; Zhou, X.; Chen, Y.; Huang, G. Adaptive Metamaterials for Broadband Sound Absorption at Low Frequencies. *Smart Mater. Struct.* **2018**, *28*, 025005. [CrossRef]
30. Berardengo, M.; Manzoni, S.; Thomas, O.; Vanali, M. Piezoelectric Resonant Shunt Enhancement by Negative Capacitances: Optimisation, Performance and Resonance Cancellation. *J. Intell. Mater. Syst. Struct.* **2018**, *29*, 2581–2606. [CrossRef]
31. Chen, Y.Y.; Huang, G.L.; Sun, C.T. Band Gap Control in an Active Elastic Metamaterial With Negative Capacitance Piezoelectric Shunting. *J. Vib. Acoust* **2014**, *136*, 061008. [CrossRef]
32. Chen, S.; Wang, G.; Song, Y. Low-Frequency Vibration Isolation in Sandwich Plates by Piezoelectric Shunting Arrays. *Smart Mater. Struct.* **2016**, *25*, 125024. [CrossRef]
33. Wang, G.; Chen, S. Large Low-Frequency Vibration Attenuation Induced by Arrays of Piezoelectric Patches Shunted with Amplifier–Resonator Feedback Circuits. *Smart Mater. Struct.* **2016**, *25*, 015004. [CrossRef]
34. Wang, G.; Cheng, J.; Chen, J. Multi-Resonant Piezoelectric Shunting Induced by Digital Controllers for Subwavelength Elastic Wave Attenuation in Smart Metamaterial. *Smart Mater. Struct.* **2017**, *26*, 025031. [CrossRef]
35. Alshaqaq, M.; Sugino, C.; Erturk, A. Digital Programming of Reciprocity Breaking in Resonant Piezoelectric Metamaterials. *Phys. Rev. Res.* **2023**, *5*, 043003. [CrossRef]
36. Li, X.; Chen, Y.; Hu, G.; Huang, G. A Self-Adaptive Metamaterial Beam with Digitally Controlled Resonators for Subwavelength Broadband Flexural Wave Attenuation. *Smart Mater. Struct.* **2018**, *27*, 045015. [CrossRef]
37. Zhang, X.; Yu, H.; He, Z.; Huang, G.; Chen, Y.; Wang, G. A Metamaterial Beam with Inverse Nonlinearity for Broadband Micro-Vibration Attenuation. *Mech. Syst. Signal Process.* **2021**, *159*, 107826. [CrossRef]
38. Luo, Z.; Chen, M.Z.; Wang, Z.X.; Zhou, L.; Li, Y.B.; Cheng, Q.; Ma, H.F.; Cui, T.J. Digital Nonlinear Metasurface with Customizable Nonreciprocity. *Adv. Funct. Mater.* **2019**, *29*, 1906635. [CrossRef]
39. Yi, K.; Matten, G.; Ouisse, M.; Sadoulet-Reboul, E.; Collet, M.; Chevallier, G. Programmable Metamaterials with Digital Synthetic Impedance Circuits for Vibration Control. *Smart Mater. Struct.* **2020**, *29*, 035005. [CrossRef]
40. Yan, B.; Wang, K.; Hu, Z.; Wu, C.; Zhang, X. Shunt Damping Vibration Control Technology: A Review. *Appl. Sci.* **2017**, *7*, 494. [CrossRef]
41. Marakakis, K.; Tairidis, G.K.; Koutsianitis, P.; Stavroulakis, G.E. Shunt Piezoelectric Systems for Noise and Vibration Control: A Review. *Front. Built Environ.* **2019**, *5*, 64. [CrossRef]
42. Zheng, Y.; Qu, Y.; Dai, S.; Chen, B.; Mao, J. Mitigating Vibration and Sound Radiation with a Digital Piezoelectric Meta-Shell in Heavy Fluids. *J. Sound Vib.* **2024**, *573*, 118221. [CrossRef]

materials

MDPI

Article

Study on Acoustic Properties of Helmholtz-Type Honeycomb Sandwich Acoustic Metamaterials

Xiao-Ling Gai *, Xian-Hui Li, Xi-Wen Guan, Tuo Xing, Ze-Nong Cai and Wen-Cheng Hu

Institute of Urban Safety and Environmental Science, Beijing Academy of Science and Technology, Beijing 100054, China; lixianh@vip.sina.com (X.-H.L.); guan0037@foxmail.com (X.-W.G.); xingtuo1991@163.com (T.X.); who518@126.com (W.-C.H.)
* Correspondence: gxldynmg@163.com

Abstract: In order to improve the acoustic performance of honeycomb sandwich structures, a Helmholtz-type honeycomb sandwich acoustic metamaterial (HHSAM) was proposed. The theoretical and finite element models were established by calculating the acoustic impedance of multiple parallel Helmholtz resonators (HR). By comparing the sound absorption of the single and multiple HR, it was found that the simulation results were basically consistent with the theoretical calculations. The sound absorption and insulation performance of the honeycomb panels, the honeycomb perforated panels, and the HHSAM structure were compared through impedance tube experiments. The results showed that, over a wide frequency range, the acoustic performance of the HHSAM structure was superior to that of the other two structures. Under scattered sound field conditions, the reverberation room results showed that the sound absorption of the HHSAM structure was better than that of the honeycomb panel in the frequency range of 100–5000 Hz. The noise reduction coefficient (NRC) of the honeycomb panel was 0.1, indicating almost no sound absorption effect in engineering. The NRC of the HHSAM structure could reach 0.35. In terms of sound insulation, the HHSAM structure was more prominent in the 400–4000 Hz range than the honeycomb panel. In the frequency range of 500–1600 Hz, the transmission loss of the HHSAM was 5 dB higher than that of the honeycomb panel.

Keywords: honeycomb structures; resonant structure; acoustic metamaterial; sound absorption; transmission loss

1. Introduction

Noise pollution mainly includes industrial noise [1], traffic noise [2], construction noise [3], and social noise [4]. With the rapid development of the economy and society, severe noise pollution has become a major concern. In industrial areas, transportation hubs, and densely populated urban areas, this noise pollution is particularly severe. Noise pollution not only disturbs the environment but also poses a major threat to public health. Noise can lead to mood disturbances, emotional instability, reduced tolerance, irritability, increased blood pressure, and long-term exposure to noise pollution, which can cause heart disease and tinnitus, thus reducing people's life expectancy [5]. Noise pollution has become a significant environmental concern in many large cities globally [6–8]. Therefore, the question of how to effectively control noise has become an important issue in the fields of environmental protection and public health.

In order to cope with noise pollution, scientists and engineers have developed various noise control methods. There are basically two methods for noise control: passive control technology [9] and active control technology [10]. A combination of these two methods can also be used. Active noise control technology can effectively suppress low-frequency noise, but its cost is high and the system is complex. The passive noise control method mainly uses techniques such as sound absorption, sound insulation, and noise reduction on the propagation path of noise to prevent the propagation of sound. Traditional sound-absorbing materials, such as porous foam and fiber materials, perform well in the middle and high frequencies but have poor acoustic performance in the low frequencies. Sound insulation technology relies on the quality and thickness of materials. To control low-frequency noise, it is necessary to use heavy and thick materials. With the increasing demand for lightweight components in aerospace, the automotive industry, and green buildings, the development of acoustic structures that combine efficient noise reduction and lightweight characteristics has become a research focus.

The honeycomb sandwich structure is widely used in aerospace, railway vehicles, ships, automobiles, construction, and other fields [11,12], because of its excellent performance, including its light weight, high specific strength, high specific stiffness, good stability, thermal insulation, and so on. Considering that the working environment in which the structure is located is often accompanied by noise and vibration, excessive noise not only causes discomfort to relevant personnel but also affects the service life of structural equipment. However, limited by the mass law, the honeycomb sandwich structure has a major disadvantage [13]. In terms of acoustic performance, compared with traditional materials such as aluminum alloys, honeycomb sandwich structures have much poorer sound insulation performance than homogeneous panels of the same quality, mainly because the coincidence frequency of sandwich panels occurs in a wider frequency range [14].

Researchers have undertaken significant investigations to boost the acoustic performance of honeycomb sandwich structures, focusing on the following key strategies: improving the stiffness of honeycomb sandwich structures [15,16]; adding porous materials [17], damping materials [18], or gas layers [19,20] inside the honeycomb; creating micropores on honeycomb sandwich structures [21]; and introducing acoustic metamaterials [13,22–25]. In addition, the negative stiffness and nonlinear effects can also dissipate a part of the system energy of the sandwich structure to reduce vibration and noise [26,27]. Most of the above studies were conducted using small samples under vertical incidence and verified through impedance tube experiments. Peiffer et al. [23] suggested that the installation conditions of the sample in the impedance tube were not consistent with the actual installation conditions of the engineering size panel. In addition, in impedance tubes, it is a plane wave, while, in engineering, it is a scattered sound field. Thus, the transmission loss (TL) of the structure measured in the impedance tube under vertical incidence cannot be directly compared with the TL of the scattered sound field environment in the field.

In order to improve the acoustic performance of honeycomb sandwich structures and provide a reference for practical engineering applications, this paper proposes a Helmholtz-type honeycomb sandwich acoustic metamaterial (HHSAM) that replaces each honeycomb in a honeycomb sandwich structure with a small Helmholtz resonator (HR), utilizing the resonance of Helmholtz resonator units and their interactions to improve the acoustic performance of the honeycomb sandwich structure. The overall structure achieves good acoustic effects while being lightweight.

2. Structure of the HHSAM

The HHSAM structure consists of a top plate, honeycomb core, and bottom plate, as shown in Figure 1. The top plate is composed of a 1-mm-thick metal aluminum plate. We produce small holes with a periodic distribution on the top plate through mechanical punching or laser punching technology. The honeycomb core is composed of resin and created through 3D printing; alternatively, an aluminum honeycomb core can be selected.

The distribution of the holes on the top plate corresponds to the position of the honeycomb core, and there is a hole on the top plate corresponding to the center of the honeycomb core. The holes on the top plate are equivalent to the neck of the Helmholtz resonator, and the cavity of the honeycomb core is equivalent to the Helmholtz resonant cavity. Each small hole and the cavity of the honeycomb core below it form a Helmholtz resonator. The bottom plate is also composed of a 1 mm aluminum plate. This structure not only has the lightweight feature of a honeycomb sandwich structure but also has the characteristics of an acoustic metamaterial.

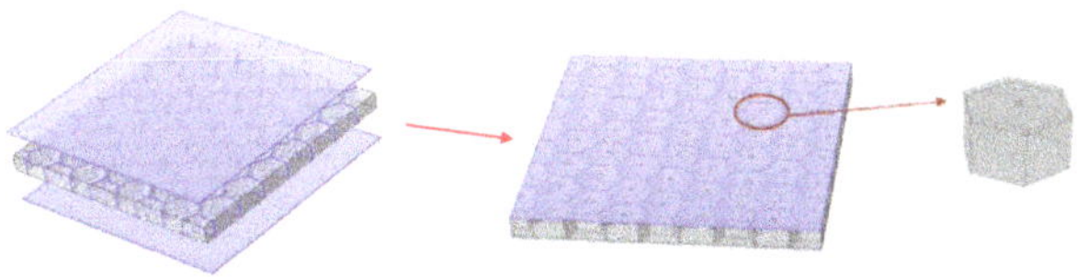

Figure 1. The Helmholtz-type honeycomb sandwich acoustic metamaterial.

3. Sound Absorption of Multiple Helmholtz Units

3.1. Sound Absorption Theory

The HHSAM structure can be seen as a parallel structure consisting of multiple Helmholtz resonators with different resonant frequencies. The sound absorption coefficient of the HHSAM structure can be obtained by calculating the acoustic impedance of the multiple Helmholtz resonators. The theoretical sound absorption coefficient α of the single Helmholtz resonator can be calculated by

$$\alpha = 1 - |R|^2. \tag{1}$$

The pressure reflection coefficient R can be estimated by

$$R = \frac{Z_{si} - 1}{Z_{si} + 1}. \tag{2}$$

Z_{si} (i = 1, 2, 3, …… n) is the normalized surface impedance of the Helmholtz resonator and can be determined by

$$Z_{si} = Z_{HRi}/\sigma, \tag{3}$$

where

$$Z_{HRi} = r_{HRi} + j\chi_{HRi}, \tag{4}$$

$$\sigma = \pi r_i^2/S, \tag{5}$$

$$S = 3\sqrt{3}/2[a + (\sqrt{3}/3)t]^2, \tag{6}$$

$$r_{HRi} = k\delta_v l'/r_i + k(\gamma - 1)\delta_t l/2/r_i, \tag{7}$$

$$\chi_{HRi} = kl'(1 + \delta_v/r_i) - \sigma cot(kD), \tag{8}$$

$$k = \omega/c_0, \tag{9}$$

$$\delta_v = \sqrt{2v/\omega}, \tag{10}$$

$$\omega = 2\pi f, \tag{11}$$

$$\delta_t \approx 0.25 \times 10^{-2}/\sqrt{f}, \tag{12}$$

$$l' = l + l_{c1} + l_{c2}, \tag{13}$$

$$l_{c1} = 0.82r_i[1 - 1.35(r_i/a) + 0.31(r_i/a)^3], \tag{14}$$

$$l_{c2} = 0.82r[1 - 0.235(r_i/a) - 1.32(r_i/a)^2 + 1.54(r_i/a)^3 - 0.86(r_i/a)^4]. \tag{15}$$

The following definitions apply.

σ: the surface porosity of the single Helmholtz resonator.
S: the single Helmholtz resonator's surface area (unit in m^2).
r_i (i = 1, 2, 3, …… n): the neck radius of the Helmholtz resonator (unit in m).
a: the edge length of the honeycomb (unit in m).
t: the honeycomb core wall thickness (unit in m).
k: the wavenumber.
δ_v: the viscous boundary layer thickness (unit in m).
ν: the kinematic viscosity of air ($\nu \approx 15 \times 10^{-6}$ m^2/s).
γ: the Poisson constant of air ($\gamma = 1.4$).
δ_t: the thermal boundary layer thickness (unit in m).
f: the frequency (unit in Hz).
l': the neck length, taking two end corrections into account (unit in m).
D: the depth of the single Helmholtz resonator (unit in m).

The normalized surface impedance of the HHSAM structure can be derived as

$$\frac{1}{Z_{ns}} = \sum_{i=1}^{n} \frac{1}{n(Z_{si})}, \tag{16}$$

and the absorption coefficient can be subsequently computed. The structural parameters of the Helmholtz resonator cell are shown in Figure 2.

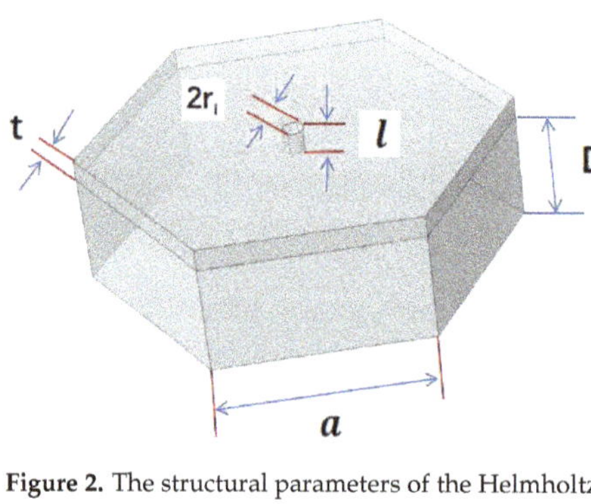

Figure 2. The structural parameters of the Helmholtz resonator cell.

3.2. Numerical Simulation

To verify the theoretical models, finite element analysis is used to calculate the sound absorption of the Helmholtz resonator cell using the COMSOL Multiphysics 5.2 software. The 3D model is used for geometric modeling. The Helmholtz resonator cell is hexagonal with a side length of 10 mm. A small hole is opened on the top surface as the neck of the Helmholtz resonator cell. Figure 3 shows the finite element model of a single Helmholtz resonator cell, double Helmholtz resonator cells, and triple Helmholtz resonator cells. In the simulations, the thermoviscous acoustics modules are used to create a calculation program. The plane wave is applied to the incidence field. The far field is a perfect matching layer that can absorb all energy. The neck length of the Helmholtz resonator cell is 1 mm. The neck diameter of the single Helmholtz resonator cell is 0.5 mm (as shown in Figure 3a). The neck diameters of the double Helmholtz resonator cells are 0.5 mm and 0.7 mm (as shown in Figure 3b). The neck diameters of the triple Helmholtz resonator cells are 0.5 mm, 0.7 mm, and 1.0 mm (as shown in Figure 3c).

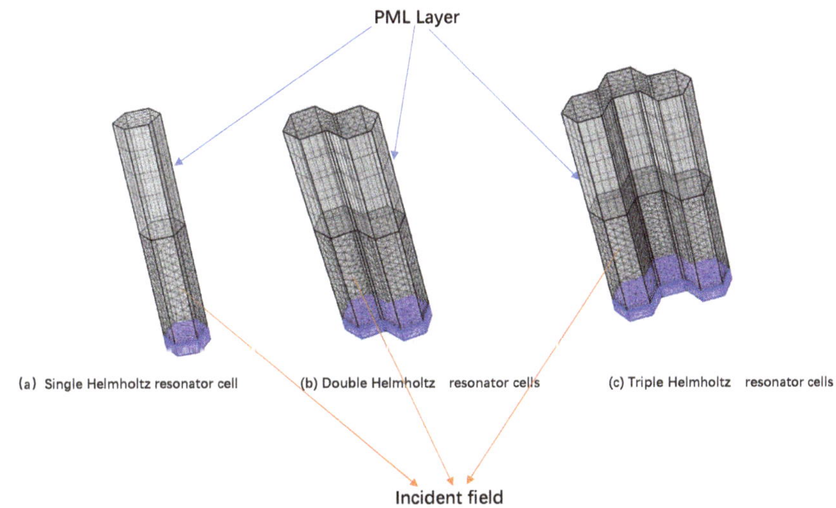

Figure 3. The finite element model of the single Helmholtz resonator cell, double Helmholtz resonator cells, and triple Helmholtz resonator cells.

Figure 4 shows the sound absorption performance curves of the single Helmholtz resonator cell, double Helmholtz resonator cells, and triple Helmholtz resonator cells obtained through theoretical and numerical simulations. The solid line represents the theoretical calculation result. The dashed line is the result of the numerical simulation. It can be seen that, regardless of whether we consider a single Helmholtz resonator cell, double Helmholtz resonator cells, or triple Helmholtz resonator cells, the theoretical calculation and numerical simulation results are consistent. Both the theoretical and numerical simulation results show that using multiple Helmholtz resonator cells can significantly broaden the sound absorption bandwidth of the structure. For example, through theoretical analysis, it can be seen that the sound absorption coefficient of a single Helmholtz resonator cell exceeds 0.5 in the frequency range of 221 Hz–356 Hz, the range of double Helmholtz resonator cells is 240 Hz–445 Hz, and the range of three Helmholtz resonator cells is expanded to 250 Hz–553 Hz. Figure 5 shows the sound pressure distribution at the absorption peak for the triple Helmholtz resonator cells. It can be seen that the three Helmholtz resonator cells play different roles at different frequencies.

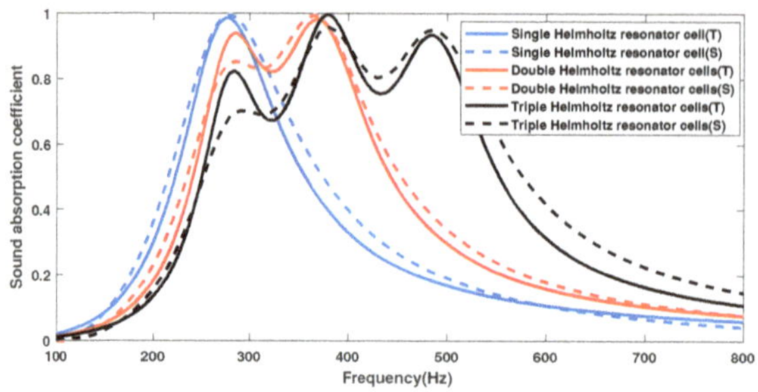

Figure 4. The sound absorption of a single Helmholtz resonator cell, double Helmholtz resonator cells, and triple Helmholtz resonator cells obtained through theoretical and numerical simulations.

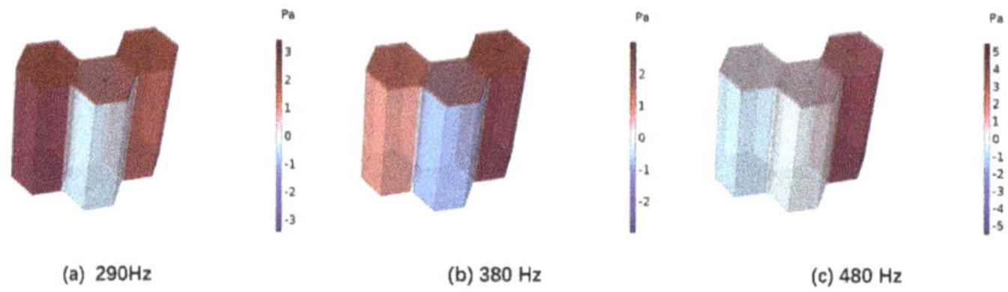

Figure 5. The sound pressure distribution of the triple Helmholtz resonator cells at three peak frequencies.

4. Experimental Study on the HHSAM Structure's Acoustic Performance

4.1. Acoustic Performance Under Vertical Incidence

4.1.1. Sound Absorption

Based on the theoretical and simulation results given in the previous section, it was found that parallel multiple Helmholtz resonators could achieve broadband sound absorption within the structure. Therefore, a HHSAM structure consisting of 19 Helmholtz resonators was designed. The HHSAM structure was constructed using parallel Helmholtz structures with different resonance frequencies, and we arranged them according to a certain periodic rule. The bottom and core layers of the HHSAM structure were created using 3D printing technology. The top panel was created by machining small holes of different diameters on a 1-mm-thick aluminum panel. Then, the two were combined through latex stations to form a HHSAM structure. The sample of the HHSAM structure is shown in Figure 6. The diameter of the sample is 100 mm. The height of the sample is 50 mm. The side length of a hexagonal honeycomb is 10 mm. The wall thickness of the honeycomb core is 2 mm. There is a hole on the top panel corresponding to each honeycomb, which represents the neck of the Helmholtz resonator. The size and distribution of the neck are shown in Figure 7 (unit in mm). A two-microphone impedance tube (type 4206) from Bruel Kjaer is applied to measure the normal incident absorption coefficient according to the standard procedure detailed in ISO (10534-2). The frequency range of measurement is from 100 to 1600 Hz. Figure 8 shows the test system of the impedance tube. Figure 9 shows the sound absorption performance of the HHSAM structure obtained through theoretical and numerical simulations and impedance tube experiments. It can be seen that the scope of action of the theoretical calculations and numerical simulations is basically consistent. The experimental results deviate slightly from the theoretical and numerical simulation results, which may be caused by manufacturing errors in the samples. Figure 10 shows the sound pressure distribution of the HHSAM structure at the peak absorption frequency. It can be seen from the distribution of the sound pressure that different Helmholtz resonator cells play different roles at different frequencies. Moreover, each Helmholtz resonance cell has a more or less mutual influence on other nearby cells. In summary, the theoretical, numerical, and experimental results effectively demonstrate that the HHSAM structure has good sound absorption performance over a wide frequency band range.

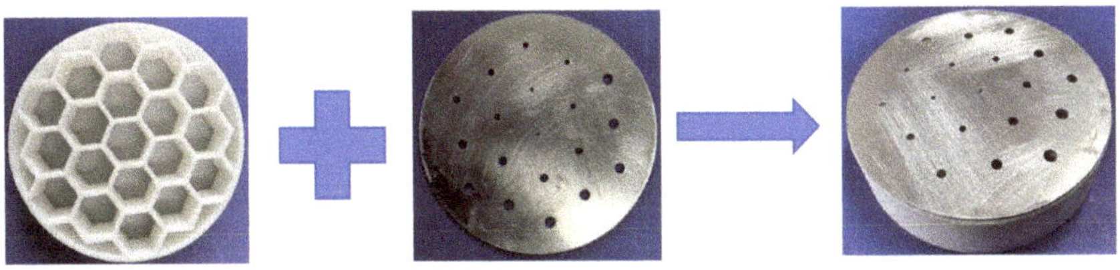

Figure 6. The sample of the HHSAM structure.

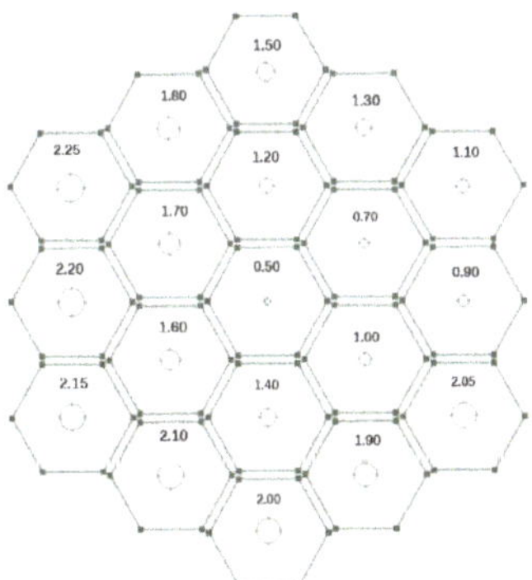

Figure 7. The neck size and distribution of the sample of the HHSAM structure.

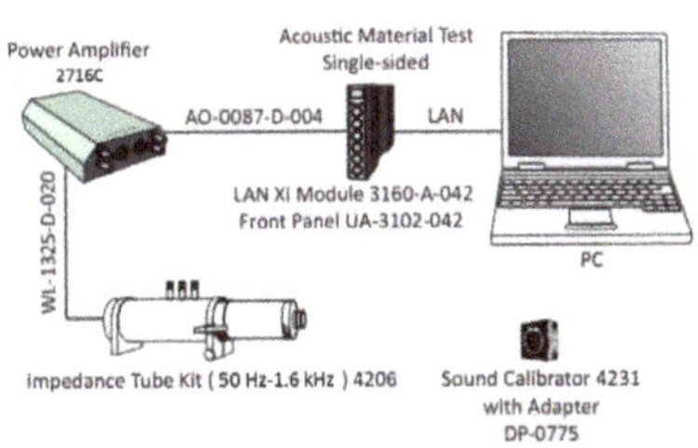

Figure 8. The test system of the impedance tube.

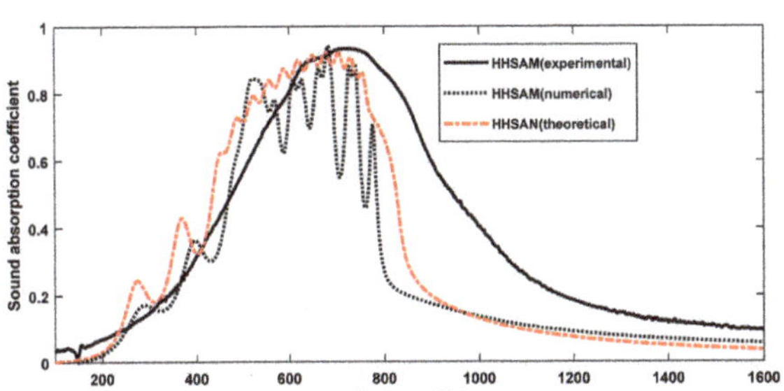

Figure 9. The theoretical, numerical, and experimental sound absorption curves of the HHSAM structure.

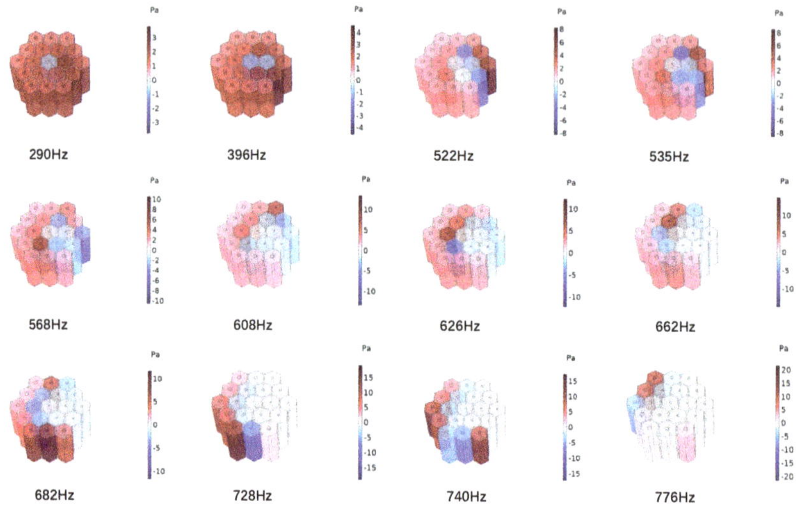

Figure 10. The sound pressure distribution of the HHSAM structure at the peak frequency.

For comparison, the acoustic performance of the honeycomb panels, the honeycomb perforated panels, and the HHSAM structure is tested separately. The thickness of all three structures is 50 mm, with only the top plate being different, as shown in Figure 11. Figure 11a shows a honeycomb panel structure, with a top plate composed of 1-mm-thick aluminum. Figure 11b shows the honeycomb perforated panel structure, with a top plate of a 1-mm-thick perforated panel and a perforated radius of 1.5 mm. Figure 11c shows the HHSAM structure, with a top plate thickness of 1 mm and 19 holes of different diameters distributed on top. Figure 12 shows the normal-incidence sound absorption coefficients of the honeycomb panel, the honeycomb perforated panel, and the HHSAM structure. It can be seen that the sound absorption performance of the honeycomb perforated panel and the HHSAM structure is significantly better than that of the honeycomb panel structure in the frequency range of 400 Hz–1600 Hz. The honeycomb perforated panel structure, with uniform holes on the upper panel, has a maximum sound absorption coefficient of 0.77 at 720 Hz, and the sound absorption coefficient is greater than 0.5 in the frequency range of 552 Hz–946 Hz. The HHSAM structure, with non-uniform holes on the upper panel, has a maximum sound absorption coefficient of 0.94 at 716 Hz, and the sound absorption coefficient is greater than 0.5 in the frequency range of 464–946 Hz. Thus, the HHSAM structure exhibits higher absorption coefficients and a wider absorption bandwidth. Figures 13 and 14 depict the measurement results regarding the normalized surface impedance of the honeycomb panel, the honeycomb perforated panel, and the HHSAM structure. The experimental results indicate that the normalized surface resistance of the honeycomb panel is too high. The surface acoustic resistance of the honeycomb perforated panel and the HHSAM structure varies around 1 in the range of 400 Hz–1600 Hz. The surface acoustic reactances of the honeycomb perforated panel and the HHSAM structure are close to 0 at the peak of sound absorption.

(a) The honeycomb panel (b) The honeycomb perforated panel (c) The HHSAM structure

Figure 11. The samples of the honeycomb panel, the honeycomb perforated panel, and the HH-SAM structure.

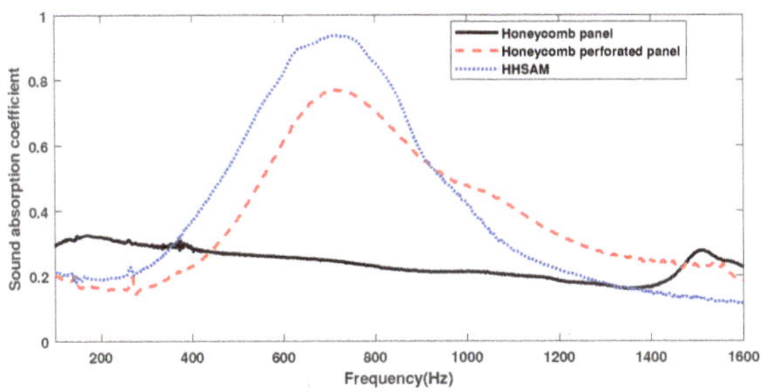

Figure 12. The experimental absorption of the honeycomb panel, the honeycomb perforated panel, and the HHSAM structure.

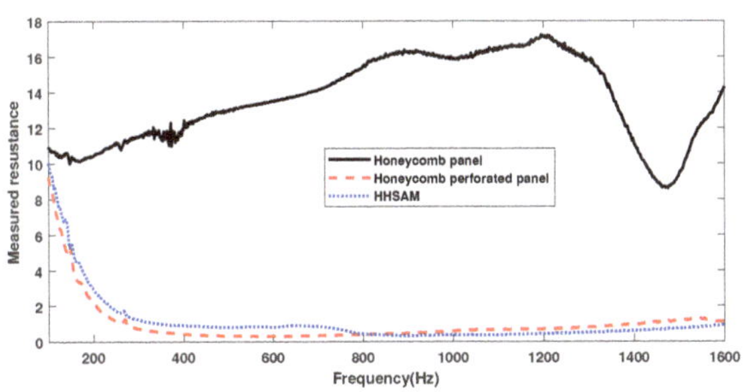

Figure 13. The measured resistance of the honeycomb panel, the honeycomb perforated panel, and the HHSAM structure.

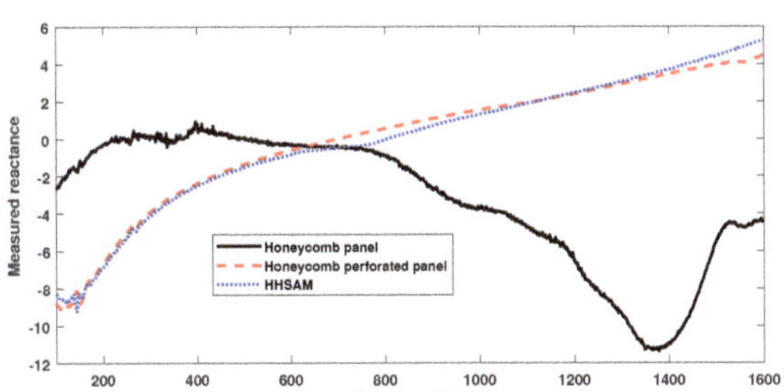

Figure 14. Themeasured reactance of the honeycomb panel, the honeycomb perforated panel, and the HHSAM structure.

4.1.2. Sound Insulation Property

Figure 15 shows the sound insulation performance of the honeycomb panel, the honeycomb perforated panel, and the HHSAM structure under vertical incidence by an impedance tube. It can be seen that, besides the narrow frequency range around 1050 Hz, the TL of the HHSAM structure in the frequency range of 100 Hz–1600 Hz is about 10 dB higher than that of the honeycomb panel structure. The TL values of the honeycomb and honeycomb perforated panels are generally similar in the range of 352 Hz–512 Hz. The TL of the honeycomb perforated panel in the ranges of 690 Hz–1040 Hz and 1160 Hz–1600 Hz is about 3 dB higher than that of the honeycomb panel structure. Overall, the TL of the HHSAM structure is higher than that of the honeycomb and honeycomb perforated panels.

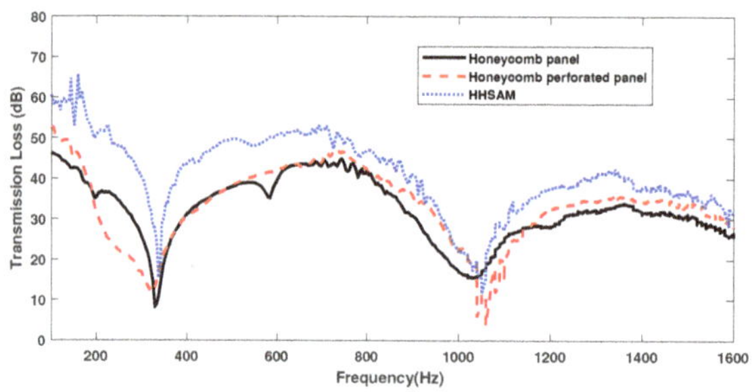

Figure 15. The TL of the honeycomb panel, the honeycomb perforated panel, and the HHSAM structure.

4.2. Acoustic Performance Under Scattered Sound Field Conditions

4.2.1. Sound Absorption

The reverberation chamber method is carried out to measure the acoustic absorption coefficients of the honeycomb panel and the HHSAM structure under scattered sound field conditions, which are close to the reality in engineering practice. The samples are measured in a 226 m^3 reverberation chamber based on ISO354. There are spherical crown diffusers at two interfaces in the reverberation chamber. The interior wall, floor, and ceiling of the reverberation chamber are all covered by smooth and hard ceramic tiles, which have strong emission effects and can be considered as having a sound absorption coefficient of 0. The equivalent sound absorption area of the test specimen A_T can be calculated using the formula

$$A_T = A_2 - A_1 = 55.3\,V\left(\frac{1}{c_2\,T_2} - \frac{1}{c_1\,T_1}\right) - 4\,V\,(m_2 - m_1) \tag{17}$$

where A_1 is the equivalent sound absorption area of the empty reverberation room,

$$A_1 = \frac{55.3\,V}{c\,T_1} - 4\,V\,m_1 \tag{18}$$

V is the volume of the empty reverberation room, c is the propagation speed of sound in air, T_1 is the reverberation time of the empty reverberation room, and m_1 is the power attenuation, calculated according to ISO9613-1 using the climatic conditions that are present in the empty reverberation room during the measurement. The value of m can be calculated from the attenuation coefficient, α, which is used in ISO9613-1 according to the formula

$$m_{1,2} = \frac{\alpha_{1,2}}{10lg(e)} \tag{19}$$

A_2 is the equivalent sound absorption area of the reverberation room containing a test specimen, which can be calculated using the formula

$$A_2 = \frac{55.3\,V}{c\,T_2} - 4\,V\,m_2 \tag{20}$$

T_2 is the reverberation time of the reverberation room after the test specimen has been introduced.

The sound absorption coefficient of a plane absorber of test objects can be calculated using the formula

$$\alpha_s = \frac{A_T}{S} \tag{21}$$

where S is the area covered by the test specimen.

The sound absorption of the honeycomb panel and the HHSAM structure is measured for 1/3 octave frequency bands from 100 Hz to 5000 Hz in the reverberation chamber. The testing instrument adopts the 2270 analyzer system, a 4190 microphone, a 2734-B power amplifier, and a 4292-L dodecahedron sound source from Denmark's B&K company (Naerum, Denmark). Figure 16 shows a schematic diagram of the reverberation chamber method used for testing. The sample of the HHSAM structure in the reverberation chamber is created by laser drilling on the upper plate of the aluminum honeycomb panel to create holes with a periodic distribution. The thickness of the honeycomb panel is 50 mm, and the thickness of the upper and lower cover panels is 1 mm. The aperture distribution is within the range of 0.5 mm–2.25 mm. The samples are placed in the middle of the reverberation chamber. Figure 17 shows the reverberation chamber experiment on the honeycomb panel and the HHSAM structure. Figure 17a shows the honeycomb panel structure, and (b) shows the HHSAM structure. The reverberation chamber's temperature is 28 °C, and the relative humidity is 55%. Figure 18 shows the sound absorption performance of the honeycomb panel and the HHSAM structure under scattered sound field conditions. The maximum sound absorption coefficient of the honeycomb panel is close to 0.2, with a noise reduction coefficient $NRC = 0.1$, indicating that it is almost impossible to use as a sound-absorbing material in engineering applications. The HHSAM structure has significantly improved sound absorption performance compared to the honeycomb panel structure. The maximum sound absorption coefficient reaches 0.55, with a noise reduction coefficient $NRC = 0.35$.

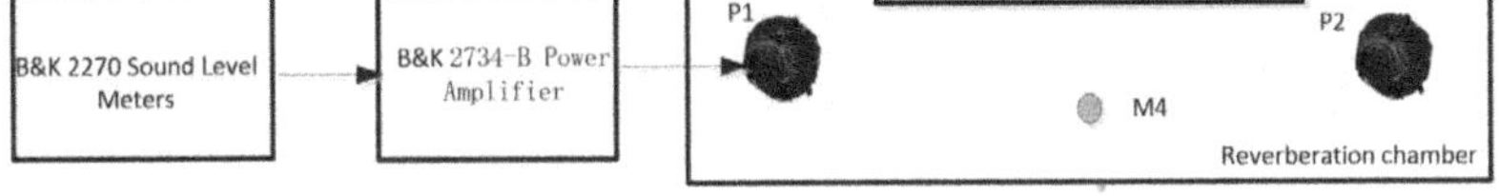

Figure 16. The schematic diagram of the reverberation chamber test.

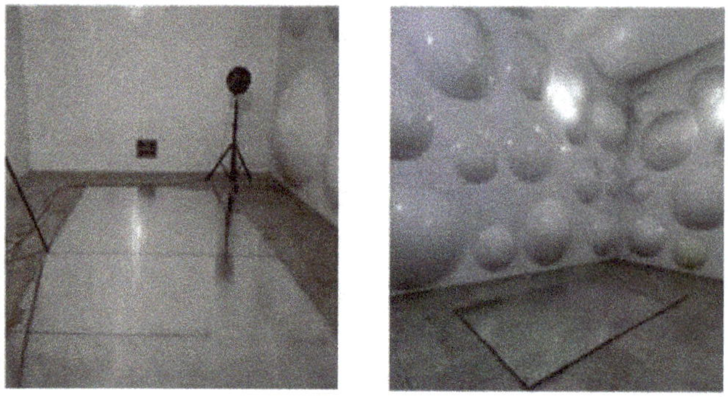

(a)The honeycomb panel (b) The HHSAM structure

Figure 17. The reverberation chamber experiment using the honeycomb panel and the HHSAM structure.

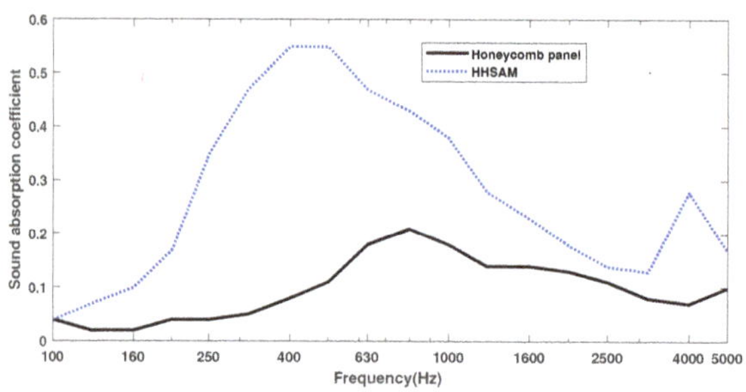

Figure 18. The sound absorption of the honeycomb panel and the HHSAM structure under scattered sound field conditions.

4.2.2. Sound Insulation Property

The sound insulation quantity of the honeycomb panel and the HHSAM structure is measured in a sound insulation laboratory. The test facility consists of two adjacent chambers with a test opening between them, in which the test specimen is inserted. The volume of the sound source chamber and reception room is 83 m^3 and 55 m^3, respectively. The area of the specimen is 2.24 m^2. The measurement is carried out in accordance with the test standards ISO 10140:2010 and ASTM E 90:2004. Two speakers are used to generate random noise in a source chamber. A half-inch microphone is used to measure the sound pressure levels at six microphone positions in each chamber. Figure 19 shows the schematic diagram of the sound isolation measurement in the laboratory. Figure 20 shows the actual sound source room (a) and the receiving room (b). Figure 21 shows the installation of the sample during the experimental process. Figure 21a is the honeycomb structure and (b) is the HHSAM structure. The insulation chamber's temperature is 26 °C, and the relative humidity is 61%. Figure 22 shows the TL characteristic of the honeycomb panel and the HHSAM structure. Compared to the honeycomb panel structure, the HHSAM structure exhibits a higher TL value in the frequency range of 400 Hz–4000 Hz. At 1250 Hz, the TL of the HHSAM structure is 8 dB higher than that of the honeycomb panel.

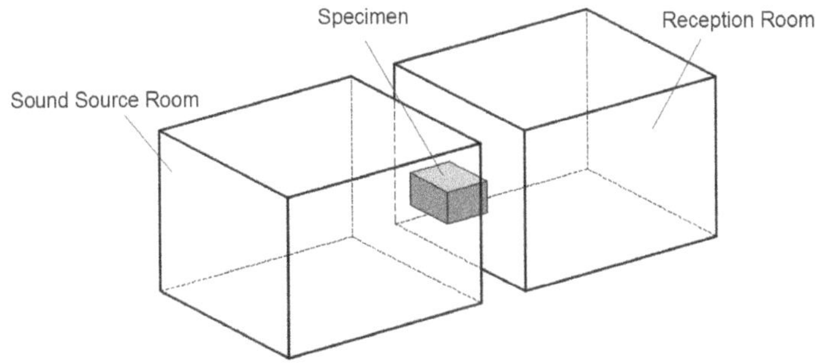

Figure 19. Schematic diagram of sound isolation measurement in the laboratory.

(a)The sound source room (b) The receiving room

Figure 20. The sound source and receiving room.

(a)The honeycomb panel (b) The HHSAM structure

Figure 21. Sample installation for the honeycomb panel and HHSAM structure.

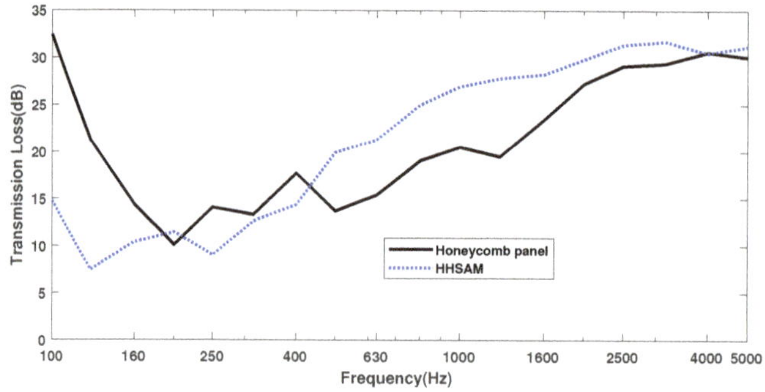

Figure 22. The TL of the honeycomb panel structure and the HHSAM structure.

5. Discussion

The research results show that honeycomb perforated panels have significantly improved acoustic performance compared to honeycomb panels. This conclusion is consistent with reference [21]. In contrast, this study further confirms this conclusion through impedance tube experiments (as shown in Figure 12). Under vertical incidence conditions, the maximum sound absorption coefficient of the honeycomb perforated panels can reach 0.77, which is much greater than the value of 0.3 for the honeycomb panel (Figure 12).

The perforation diameter of each structure in reference [21] is consistent, and their research mainly focuses on the influence of different types of honeycomb cores (hexagonal, rectangular, triangular) and the impact of changes in perforation on the acoustic performance of the structure. In contrast, in this study, we used different apertures with periodic distributions, theoretically forming more absorption peaks through different Helmholtz resonances, which could further expand the absorption bandwidth of the structure. We also verified this through experiments (as shown in Figure 12). The maximum sound absorption coefficient of the HHSAM structure can reach 0.94, which is higher than the 0.77 of the honeycomb perforated panel. The absorption bandwidth of the HHSAM structure with a sound absorption coefficient greater than 0.5 is also extended towards lower frequencies by about 100 Hz compared to the honeycomb panel structure. Compared with the 17-mm-thick hexagonal honeycomb core sandwich panel with perforated faceplates used in reference [21], we also theoretically calculated the sound absorption of a 17-mm-thick HHSAM structure, as shown in Figure 23. The perforation diameter in reference [21] was 1mm, while the perforation diameter of the HHSAM structure was between 1.0 mm and 2.7 mm. It can be seen that, under the same thickness, the sandwich panel with perforated faceplates reaches a maximum sound absorption coefficient of 0.82 at 840 Hz, and the sound absorption coefficient is greater than 0.5 in the range of 730 Hz–980 Hz. The maximum sound absorption coefficient of the HHSAM structure can reach 0.97 at 1002 Hz, and the sound absorption coefficient is greater than 0.5 in the range of 540 Hz–1109 Hz. This shows that using structures with different apertures with a periodic distribution can better improve the acoustic performance of honeycomb panel structures than using a single-aperture structure.

Unlike the existing literature, in order to facilitate engineering applications, this study produced a large sample of 2.24 m^2 and conducted experiments with irregular incidence. The results also showed that the HHSAM structure had better acoustic performance than honeycomb panels (Figures 18 and 22). In terms of the production process, only the top plate of the original honeycomb panel is replaced with a perforated panel with periodic distribution holes. This indicates that mass production is achievable.

Although these results are promising, there are still some challenges. To improve the acoustic performance of the HHSAM structure, it is necessary to further optimize the structural parameters based on the operating frequency. The impact of oblique incidence on the acoustic performance of HHSAM structures needs further analysis, and the influence of opening holes on the strength of honeycomb panels on the top surface also needs experimental evaluation.

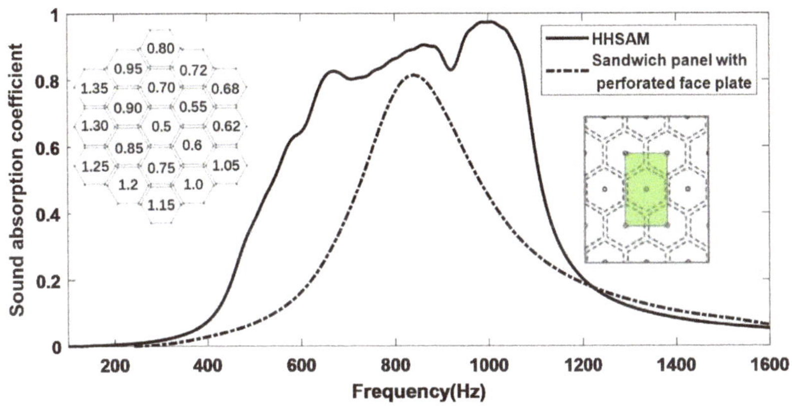

Figure 23. The acoustic performance of a 17-mm-thick sandwich panel with perforated faceplates and the HHSAM structure.

6. Conclusions

In this study, the acoustic properties of the HHSAM structure were investigated in terms of theory, simulations, and experiments. The theoretical calculations were validated by comparing them with numerical results, and excellent agreement was achieved. The theoretical, simulation, and experimental results all demonstrate that a wide frequency range of sound absorption can be achieved by the parallel connection of multiple Helmholtz resonators with different resonant frequencies. The impedance tube experiments showed that the maximum sound absorption coefficient of the HHSAM structure could reach 0.9, which was much greater than the 0.3 of the honeycomb panel. At the same time, in the frequency range of 100 Hz–1600 Hz, the TL of the the HHSAM structure was 10 dB higher than that of the honeycomb panel structure. The honeycomb perforated panels also could not achieve this effect. Under scattered sound field conditions, the sound absorption and insulation performance of the HHSAM structure was also better than that of the honeycomb panel structure over a wide frequency band range. In the frequency range of 428 Hz to 3906 Hz, the TL of the HHSAM structure was significantly higher than that of honeycomb panels. At 1250 Hz, the TL of the HHSAM structure was 8 dB higher than that of honeycomb panels.

Author Contributions: Methodology, X.-L.G., X.-H.L. and W.-C.H.; Software, X.-L.G., X.-W.G. and T.X.; Formal analysis, X.-L.G.; Investigation, Z.-N.C.; Writing—original draft, X.-L.G. All authors have read and agreed to the published version of the manuscript.

Funding: This work was supported by the Beijing Natural Science Foundation, No. 1222021; the National Natural Science Foundation of China, Grant No. 12274038; the Financial Program of BJAST, No. 24CB001-11 and 25CB001-08; and the Taishan Industrial Experts Program.

Institutional Review Board Statement: Not application.

Informed Consent Statement: Not application.

Data Availability Statement: The original contributions presented in this study are included in the article. Further inquiries can be directed to the corresponding author.

Conflicts of Interest: The authors declared no potential conflicts of interest.

References

1. Wang, Y.B.; Xie, J.L.; Lee, H.M.; Lee, H.P. Industrial noise reduction measures based on the Italian prioritisation index. *J. Build. Eng.* **2025**, *99*, 111642. [CrossRef]
2. Danilevicius, A.; Karpenko, M.; Krivanek, V. Research on the noise pollution from different vehicle categories in the urban area. *Transport* **2023**, *38*, 1–11.
3. Wu, Z.F.; Zhao, X.Q. Reducing construction noise: Sound masking effect on soundscape dominated by construction noise. *Int. J. Environ. Sci. Technol.* **2025**, *22*, 797–832.
4. Yang, D.L.; Liu, X.H.; Ren, Z.C.; Li, M.N. Relation between noise pollution and life satisfaction based on the 2019 China social survey. *Int. J. Environ. Res. Public Health* **2022**, *19*, 7015.
5. Kumar, S.; Xiang, T.B.; Lee, H.P. Ventilated acoustic metamaterial window panels for simultaneous noise shielding and air circulation. *Appl. Acoust.* **2020**, *159*, 107088. [CrossRef]
6. Quader, M.A.; Rahman, M.M.; Chisty, M.A.; Hattawi, K.S.A.; Islam, E.A.M.K. Evaluation of noise pollution impact on health in Dhaka city, Bangladesh. *Front. Public Health* **2024**, *12*, 2296–2565.
7. Monazzam, M.; Karimi, E.; Abbaspour, M.; Nassiri, P.; Taghavi, L. Spatial traffic noise pollution assessment—A case study. *Int. J. Occup. Med. Environ. Health* **2015**, *28*, 625–634.
8. Stanovsk, M.; Tomskov, H.; Slachtov, H.; Potuznkov, D.; Argalsov, L. Health impact of environmental and industrial noise—A narrative review. *Med. Pr.* **2024**, *75*, 425–431.
9. Stosiak, M.; Yatskiv, I.; Prentkovskis, O.; Karpenko, M. Reduction of pressure pulsations over a wide frequency range in hydrostatic system. *Machines* **2025**, *13*, 25. [CrossRef]
10. Lee, H.M.; Hua, Y.T.; Wang, Z.M.; Lim, K.M.; Lee, H.P. A review of the application of active noise control technologies on windoes: Challenges and limitations. *Appl. Acoust.* **2021**, *174*, 107753.
11. He, M.F.; Hu, W.B. A study on composite honeycomb sandwich panel structure. *Mater. Des.* **2008**, *29*, 709–713. [CrossRef]
12. Zheng, X.; Lee, H.; Weisgraber, T.H.; Shusteff, M.; DeOtte, J.; Duoss, E.B. Ultralight, ultrastiff mechanical metamaterials. *Science* **2014**, *344*, 1373–1377. [PubMed]
13. Li, Y.L.; Zhang, Y.L.; Xie, S. A lightweight multilayer honeycomb membrane-type acoustic metamaterial. *Appl. Acoust.* **2020**, *168*, 107427.
14. Huang, W.C.; Ng, C.F. Sound insulation improvement using honeycomb sandwich panels. *Appl. Acoust.* **1998**, *53*, 163–177.
15. Ng, C.F.; Hui, C.K. Low frequency sound insulation using stiffness control with honeycomb panels. *Appl. Acoust.* **2008**, *69*, 293–301.
16. Kang, L.D.; Liu, B.L.; An, F.Y.; Wei, D.P. Mechanical and vibro-acoustic performance of sandwich panel with perforated honeycomb cores. *J. Acoust. Soc. Am.* **2022**, *152*, 1539–1546.
17. Yang, Y.; Li, B.B.; Chen, Z.F.; Sui, N.; Chen, Z.; Saeed, M.U.; Li, Y.F.; Fu, R.; Wu, C.; Jing, Y. Acoustic properties of glass fiber assembly-filled honeycomb sandwich panels. *Compos. Part B* **2016**, *96*, 281–286.
18. Boztoprak, Y.; Ünal, M.; Özada, Ç.; Kuzu, E.; Özer, H.; Ergin, F.; Yazıcı, M. Sound insulation performance of honeycomb core aluminum sandwich panels with flexible epoxy-based foam infill. *Compos. Struct.* **2023**, *319*, 117149.
19. Rajaram, S.; Nutt, S. Measurement of sound transmission losses of honeycomb partitions with added gas layers. *Noise Control Eng. J.* **2006**, *54*, 101–105.
20. Naify, C.J.; Huang, C.Z.; Sneddon, M.; Nutt, S. Transmission loss of honeycomb sandwich structures with attached gas layers. *Appl. Acoust.* **2011**, *72*, 71–77.
21. Meng, H.; Galland, M.A.; Ichchou, M.; Xin, F.X.; Lu, T.J. On the low frequency acoustic properties of novel multifunctional honeycomb sandwich panels with micro-perforated faceplates. *Appl. Acoust.* **2019**, *152*, 31–40. [CrossRef]
22. Peng, X.Y.; Ji, J.; Jing Y. Composite honeycomb metasurface panel for broadband sound absorption. *J. Acoust. Soc. Am.* **2018**, *144*, 255–261. [CrossRef] [PubMed]
23. Sui, N.; Yan, X.; Huang, T.Y.; Xu, J.; Yuan, F.G.; Jing, Y.A. Lightweight yet sound-proof honeycomb acoustic metamaterial. *Appl. Phys. Lett.* **2015**, *106*, 171905. [CrossRef]
24. Lu, K.; Wu, J.H.; Guan, D.; Gao, N.S.; Jing, L. A lightweight low-frequency sound insulation membrane-type acoustic metamaterial. *AIP Adv.* **2016**, *6*, 025116. [CrossRef]
25. Peiffer, A.; Grunewald, M.; Lempereur, P.; Ni, S.; Xiang, Y.; Huang, T.Y.; Xu, J.; Yuan, F.G. Comment on "A lightweight yet sound-proof honeycomb acoustic metamaterial". *Appl. Phys. Lett.* **2015**, *107*, 216101. [CrossRef]

26. Meng, H.; Huang, X.C.; Chen, Y.Y.; Theodossiades, S.; Chronopoulos, D. Structural vibration absorption in multilayered sandwich structures using negative stiffness nonlinear oscillators. *Appl. Acoust.* **2021**, *182*, 108240. [CrossRef]
27. Yasser, D.; Mohamed, A.E. Numerical and experimental investigation of negative stiffness beams and honeycomb structures. *Eng. Struct.* **2024**, *301*, 117163.

materials

MDPI

Article

Theoretical Estimation Based on CT Images and Experiments on the Sound Absorption Coefficient of Foam Materials with Removed Membrane †

Shuichi Sakamoto [1,*], Takamasa Satoh [2], Kaito Tanabe [3], Koki Maruyama [3] and Yusei Himori [3]

[1] Department of Engineering, Niigata University, Ikarashi 2-no-cho 8050, Nishi-ku, Niigata City 950-2181, Niigata Prefecture, Japan
[2] Fukoku Co., Ltd., 6 Showa Chiyoda-machi, Oura-gun 370-0723, Gunma Prefecture, Japan; takamasa_satoh@fukoku-rubber.co.jp
[3] Graduate School of Science and Technology, Niigata University, Ikarashi 2-no-cho 8050, Nishi-ku, Niigata City 950-2181, Niigata Prefecture, Japan
* Correspondence: sakamoto@eng.niigata-u.ac.jp; Tel.: +81-25-262-7003
† Presented at 17th International Conference on Motion and Vibration Control and the 20th Asia-Pacific Vibration Conference (MoViC&APVC2024), Tokyo, Japan, 5–8 August 2024.

Abstract: The structure of foam sound absorbers is not strictly regular, and it is difficult to create a geometric model. In this study, a method for estimating the sound absorption properties of foam sound absorbers with the membrane removed was proposed based on computed tomography (CT) scan images: the circumference of the structure and the cross-sectional area of the voids in the foam cross-section were determined from CT scans of foam materials. The propagation constant and characteristic impedance at the voids were obtained by approximating the foam material cross-section as the clearance between two planes, and the transfer matrix method was used to calculate the normal incident sound absorption coefficient. Further, the sound absorption coefficient was theoretically derived using the effective density with the measured tortuosity applied and compared with the experimental value using a two-microphone impedance measuring tube. By extracting the skeletal part of foam materials by using image processing and removing the residual noise in CT images, and then varying the correction factor for the skeleton surface area, the theoretical value of the sound-absorbing foam without a membrane was closer to the measured value.

Keywords: foam absorbent; micro-CT scan images; tortuosity; sound absorption coefficient; transfer matrix method

1. Introduction

Recently, with the electrification of mechanical products, high-frequency noise due to magnetostrictive noise from motors caused by inverters has become apparent [1]. In addition, thin, lightweight sound-absorbing materials that can be used in various shapes are required for noise reduction in moving vehicles, such as automobiles [2,3]. Research on porous materials has focused on evaluating and optimizing their acoustic and mechanical properties, with applications such as road surfacing [4,5]. In addition, studies on recycled porous materials, like polyurethane foams, have highlighted their potential as sound-absorbing materials [6,7].

In this study, we focused on polyurethane foam, which is a porous sound-absorbing material. Polyurethane foams are classified according to their hardness as hard, semi-hard, or soft, and each type has different application areas and properties. Polyurethane foams are used in car dashboards [8] and seats [9], and research [10] has been conducted to recycle polyurethane foam waste from the textile industry. The polyurethane foams used in these applications are soft foams manufactured by mixing a foaming agent. The cell size and membrane proportion vary depending on the manufacturing technique [11,12].

The properties of foam sound-absorbing materials vary greatly depending on the porosity and cell size [12]. Observation of foam sound-absorbing materials via electron microscopy reveals the presence of a membrane on the foam material framework. The membranes of foam materials have not been described in previous studies on foam sound-absorbing materials [13,14]. Meanwhile, experimental studies on the presence or absence of membranes in foam materials have reported that the influence of membranes on the sound absorption coefficient is significant [15,16]. The Kelvin model is a well-known, homogeneous mathematical model for foam materials. However, real foam materials exhibit irregular structures, as revealed by micrographs and CT images. The structure of foam sound absorbers is not strictly regular, and it is difficult to create a geometric model.

Therefore, we theoretically estimate the acoustic properties of foam sound absorbers. As a basic study, a preliminary study was conducted on a sample with its membrane removed to investigate the sound absorption properties of foam sound absorbers in the absence of membranes [17].

In this study, a method for estimating the sound absorption properties of foam sound absorbers with the membrane removed was proposed based on computed tomography (CT) scan images: the circumference of the structure and the cross-sectional area of the voids in the foam cross-section were determined from CT scans of foam materials. The propagation constant and characteristic impedance at the voids were obtained by approximating the foam material cross-section as the clearance between two planes, and the transfer matrix method was used to calculate the normal incident sound absorption coefficient. Further, the sound absorption coefficient was theoretically derived using the effective density with the measured tortuosity applied and compared with the experimental value using a two-microphone impedance measuring tube.

This method for estimating the sound absorption coefficient using CT scan images has yielded positive results for other materials, such as randomly filled granular materials [18], rice and buckwheat husks [19], and rice straw [20]. This study explores the potential for developing image-based models using tomographic images of foam materials, as opposed to the mathematical models typically used for grid structures [21] and frame structures [16] in previous studies. This study is expected to be useful for developing foam sound absorbers based on simulation.

2. Sample and Measuring Apparatus

2.1. Sample Used for Measurements

A porous foam sound-absorbing material was used in this study. The material was L-25 manufactured by Toyo Quality One (Kawagoe, Saitama, Japan), and samples with a 29 mm diameter and 10 mm height were prepared.

A scanning electron microscope (SEM; JEOL JSM-6010PLUS/LA, Tokyo, Japan) confirmed the structures and films (Figure 1a,b). According to a previous study [16], the percentage of grids with complete membranes was 28% of the total. Figure 1a,b show different magnification scales. In Figure 1b, the areas circled in red and blue are the rod cross-section and membrane, respectively.

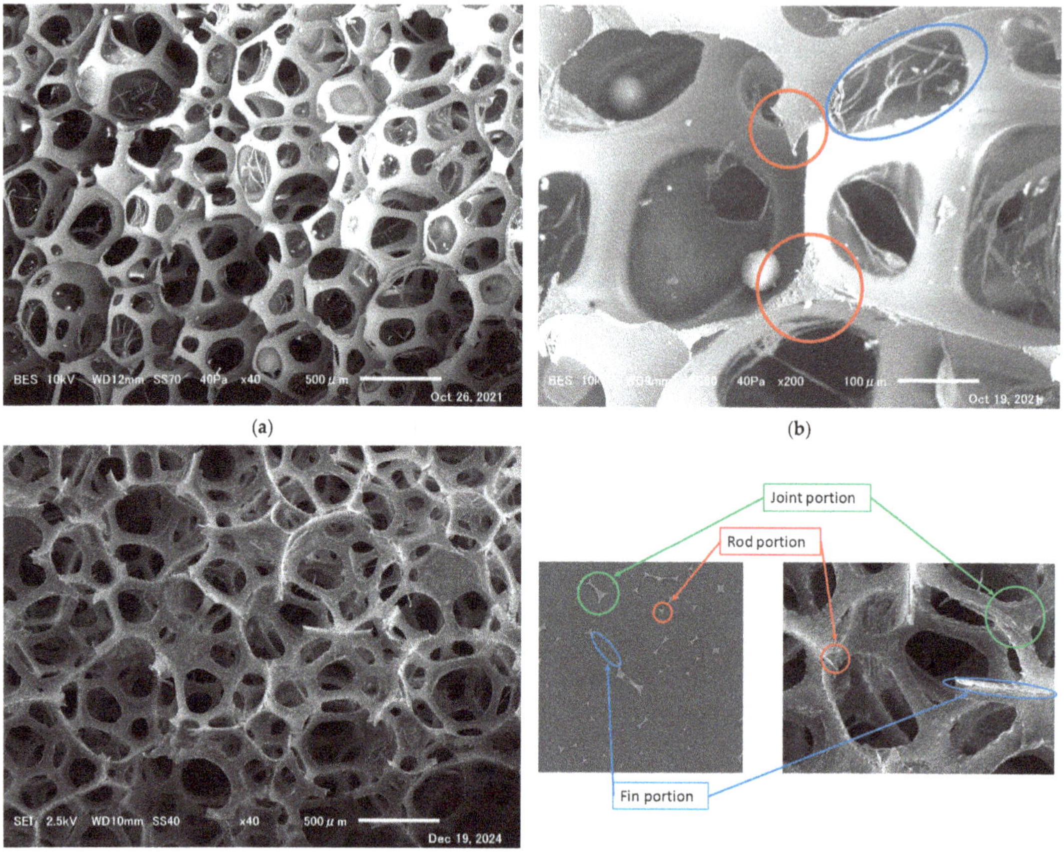

(**a**) (**b**)

(**c**) (**d**)

Figure 1. Magnified electron micrograph of the foam sound-absorbing material and the names of its parts (JEOL Ltd. JSM-6010PLUS/LA): (**a**) structure of the foam material and membrane presence, (**b**) membrane (blue circle) and cross-section of the rod portion (red circles), (**c**) image of the foam material with the membrane removed, and (**d**) names of the parts of the framework.

Notably, membranes present in sound-absorbing foam materials are as thin as 1.5 ± 0.25 μm [12] and difficult to completely image with the resolution of the micro–X-ray CT used (3.9 μm). Therefore, a comparison was made between the experimental and theoretical values based on CT images for the foam materials with the membranes removed. Foam materials without membranes are known for usage in microphone wind screens and water absorption; however, their sound absorption performance is very poor. As this research concerned the fundamentals of sound-absorbing foams, it dealt with the foam skeleton of sound-absorbing foams. The technique used to remove membranes from foam sound-absorbing materials was rubbing the foam material dozens of times in water, thereby using hydraulic pressure to break the membranes.

Figure 1c shows an SEM image of a foam sound absorber with its membrane removed, at the same magnification as in Figure 1a. Figure 1c shows that the membrane was almost completely removed.

The names of the parts of the foam sound absorber framework are shown in Figure 1d. The rod-like part (like a triangular pillar) is called the rod (portion), the connecting part between the rods is called the joint (portion), and the thin-walled part around the rod or joint that connects to the membrane is called the fin portion.

Figure 2 shows a photograph of the sample used to measure the sound absorption coefficient.

Figure 2. Sample used for impedance tube measurements.

2.2. *Equipment for Measuring the Sound Absorption Coefficient*

A two-microphone acoustic impedance tube (Type 4206 from Brüel & Kjær, Nærum, Denmark) as shown in Figure 3 was used to measure the normal incident sound absorption coefficient. The transfer function between the sound pressure signals from the two microphones was measured using an FFT analyzer. This transfer function was then used to calculate the sound absorption coefficient at normal incidence, following the guidelines of ISO 10534-2 [22]. The sample used in this study did not have a high sound absorption coefficient in the low-frequency range; thus, an impedance tube with an inner diameter of 29 mm was used.

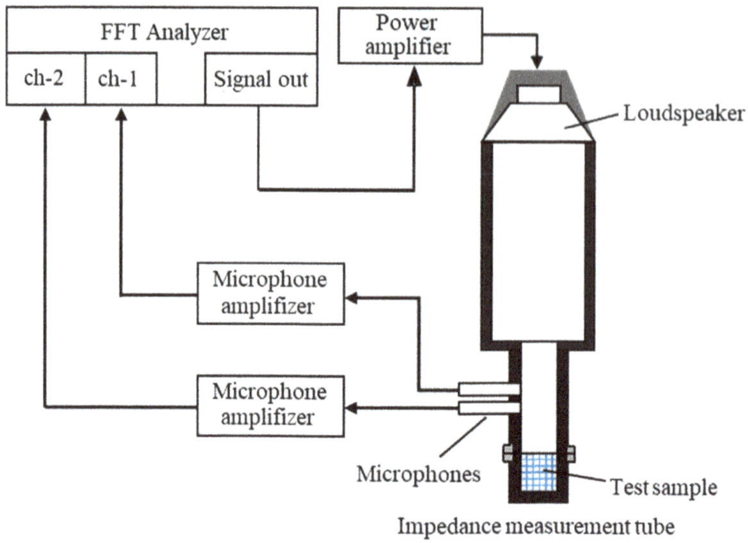

Figure 3. Schematic of two-microphone impedance tube used for absorption coefficient measurement.

2.3. Methods and Results of Measuring Tortuosity

When a material's internal geometry is complex, sound waves pass along a path longer than the material's length in the travel direction. Tortuosity is a parameter that expresses the complexity of sound waves as they pass through a material. In this study, the tortuosity of foam sound absorbers was measured using ultrasonic sensors.

A schematic of the tortuosity measurement system is shown in Figure 4. A sample holder was fabricated using light-curing resin with a Form2 3D printer (manufactured by Formlabs, Somerville, MA, USA) to secure the foam material. The sample holder used for measuring the tortuosity and a photograph of the sample are shown in Figure 5. The same method previously reported [18,19,23] was used to derive the tortuosity.

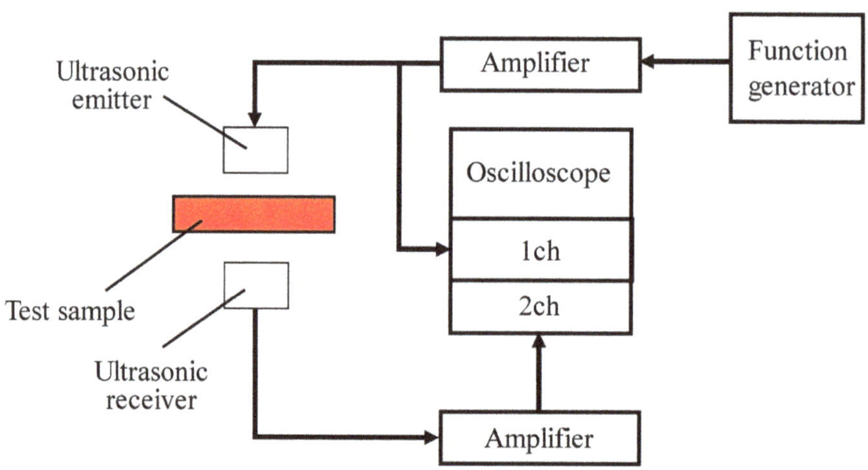

Figure 4. Schematic of tortuosity measurement.

The tortuosity measurement results are shown in Figure 6. Ultrasonic sensors with central frequencies of 58, 110, 150, 200, and 300 kHz were used. In Figure 6, the horizontal axis represents the inverse of the square root of the frequency used for measurement, while the vertical axis plots the tortuosity for each frequency, as determined by the measurements. The intercept of the least-squares line in Figure 6 gives a tortuosity of $\alpha_\infty = 1.26$ for the sound-absorbing foam sample without a membrane. The result for the foam sound absorber before the membrane was removed is also shown in Figure 6 for reference, where the tortuosity was determined to be $\alpha_\infty = 1.95$.

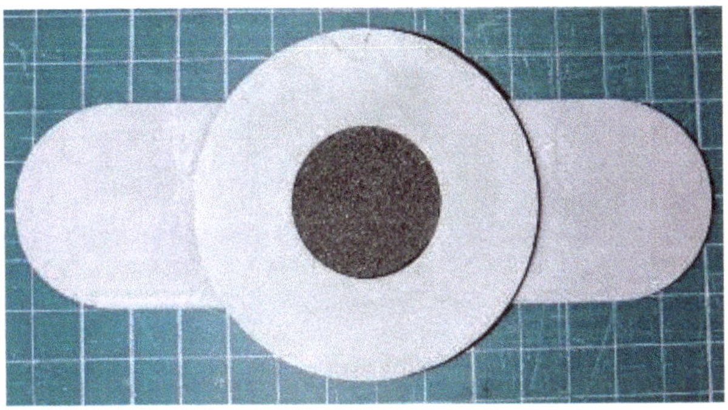

Figure 5. Test sample for tortuosity measurement with a foam sound absorber.

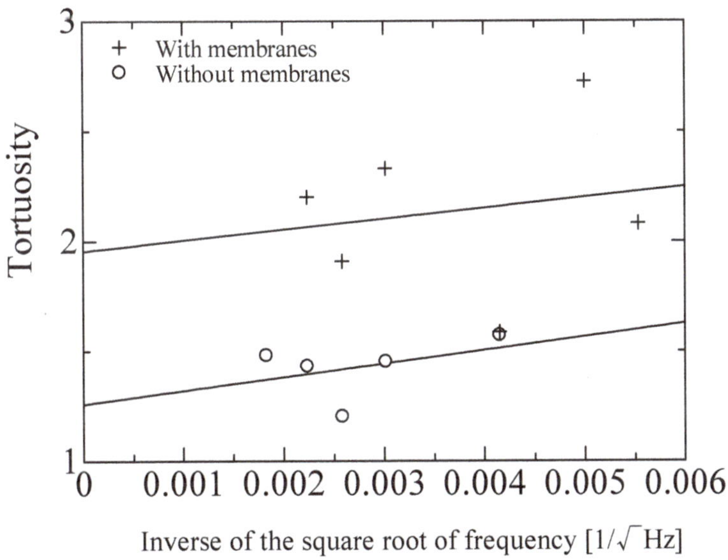

Figure 6. Tortuosity measurement results (foam sound absorber with or without membrane).

3. Theoretical Analysis

3.1. CT Scan Tomograms

An example of a CT scan tomogram of a foam material without a membrane is shown in Figure 7a. The equipment used was an MCT225 Metrology CT from NIKON Corp. (Tokyo, Japan).

As shown schematically in Figure 7b, the tomograms were taken in the y-z plane, which is perpendicular to the sound wave incidence direction (x-direction). The area of the images is a square with a side of approximately 4.0 mm. The number of images used for the theoretical analysis was 1,026. Therefore, the pitch of the images in the x-direction is approximately 3.9 μm. And the resolution (pixel size) of the image is 3.9 μm square.

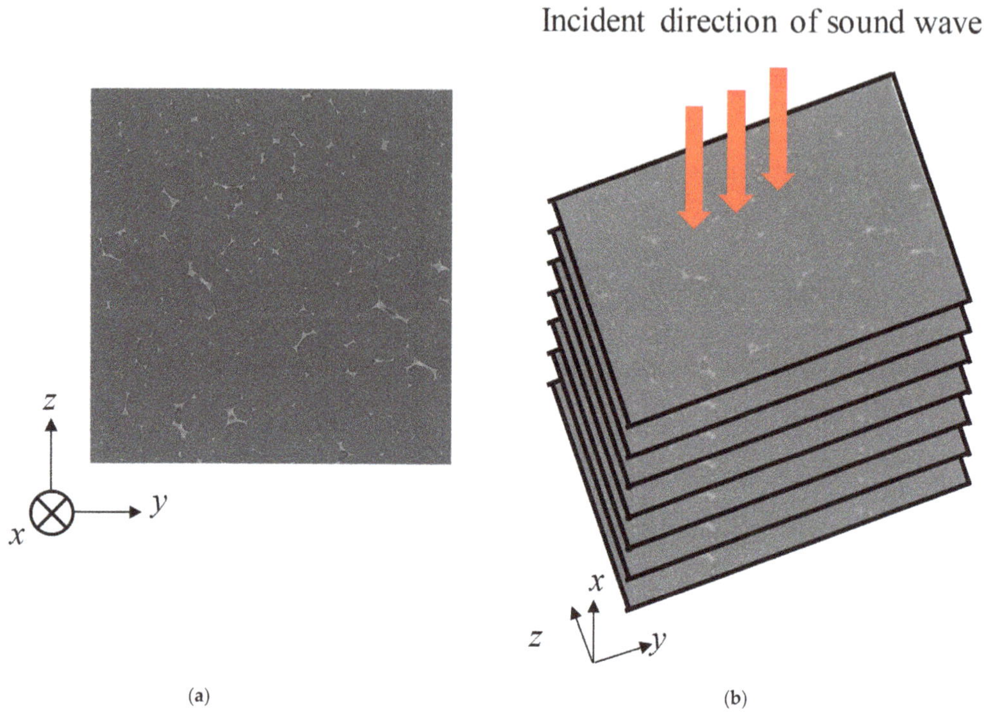

Figure 7. Cross-sectional image of foam sound absorber: (**a**) typical cross-sectional image of an arbitrary point in the x-direction without membrane and (**b**) schematic of analysis units.

3.2. CT Images and Image Processing

The internal structure of foam materials is complex (Figure 1a), making it difficult to create analogous mathematical models. Therefore, the sound absorption coefficient was theoretically estimated from the actual structure of foam materials by layering the cross-sectional images using CT.

The CT images were binarized to determine the cross-sectional area of voids. The CT images used were in 8-bit (256 grayscale). They were converted into black-and-white binarized images using a set "threshold" as a boundary. An under- or over-threshold can increase noise in a binarized image and cause the loss of details (e.g., fin portions) observed in the original CT image. Thus, the original CT and binarized images were compared, and the threshold was varied so that the fin portion could be extracted and denoised appropriately.

The amount of noise and the extraction performance of the fin portion differ depending on the spatial resolution of the CT scanner used and the range of the acquisition. Therefore, noise reduction and extraction of the form skeleton were performed in accordance with the tomographic images of the X-ray CT equipment used in this study.

The processing of the tomograms taken by the NIKON Corp. MCT225 Metrology CT, as shown in Figure 7a, is described. The tomograms contain noise. Therefore, if binarization is performed before denoising, excessive noise will remain if a threshold value that allows the fin portion to be extracted is selected. Thus, a general median filter was used for image noise removal with a kernel size of 3.

Figure 8a–c show the binarization results of the X-ray CT scan images after the preliminary processing. The threshold values in Figure 8a–c differ. The cross-sectional area of the voids is determined by the number of pixels in the area identified as black in the images [19].

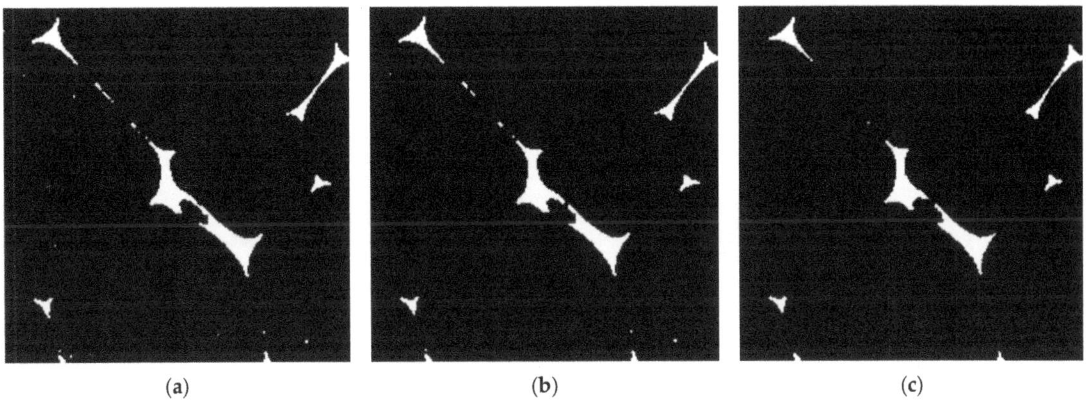

Figure 8. Part of binarized images in Figure 7a scanned by MCT225 Metrology CT (enlarged image of an area of approximately 0.90×0.90 mm^2) with thresholds of (**a**) 0.499, (**b**) 0.500, and (**c**) 0.520.

The Canny edge detection method [24] was applied to the denoised binarized images for edge extraction. This helped sharpen the structural outline and obtain a circumference closer to the actual structure. This gave the circumference of the skeletal part [19].

3.3. Approximation of Void to Clearance Between Two Planes

This section describes the approximation of a void to the clearance between the two planes. The cross-sectional area and circumference of voids were obtained from the binarized and edge-extracted images to approximate the clearance between the two planes. Figure 9a,b show the geometry before and after the approximation to a clearance between two planes, respectively, which are calculated using the cross-sectional area and circumference of the gap obtained in Section 3.2. Multiplying the cross-sectional area of the voids by the pitch d between the images yields the volume of the voids, V_n, shown in Figure 9a. Similarly, multiplying the total circumference of the cross-section by the pitch d yields the surface area of the skeletal part, S_n, in Figure 9a. The clearance thickness b_n between the two planes is determined for each image shown in Figure 9b using the relationship between the volume V_n and the surface area S_n, as described in Equation (1):

$$b_n = \frac{2V_n}{S_n \times F} \quad (1)$$

where F is a correction factor to account for the true surface area and represents the ratio of the real surface area to S_n. For example, if the foam material skeleton is assumed to be the area surrounded by three circumscribed circles (Figure 10a), the correction factor F for obtaining the true surface area is $\pi/2 \cong 1.57$. If the skeleton is assumed to be a flat plate inclined at 45° in the x-direction (Figure 10b), the correction factor F is $\sqrt{2} \cong 1.414$.

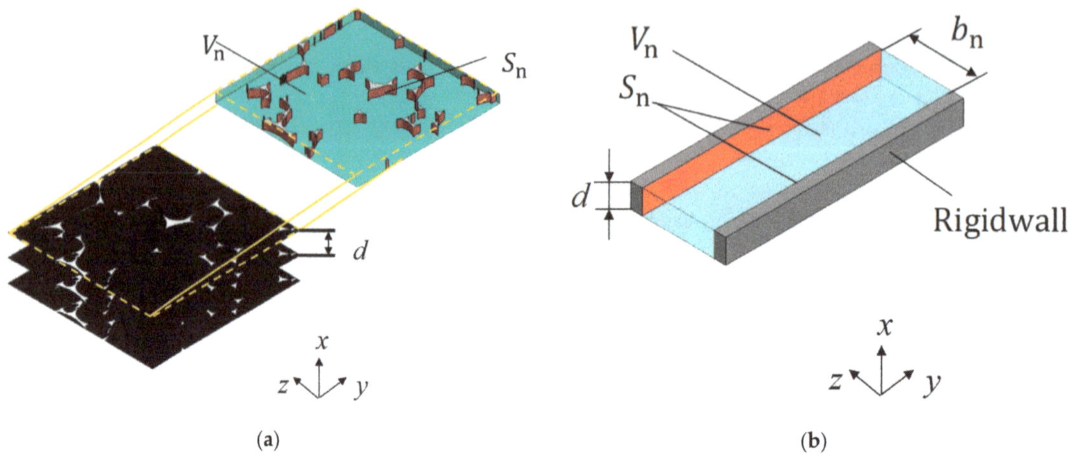

Figure 9. Foam surface area and clearance volume: (**a**) cross-sectional image of an arbitrary point in the x-direction and (**b**) two-plane approximation.

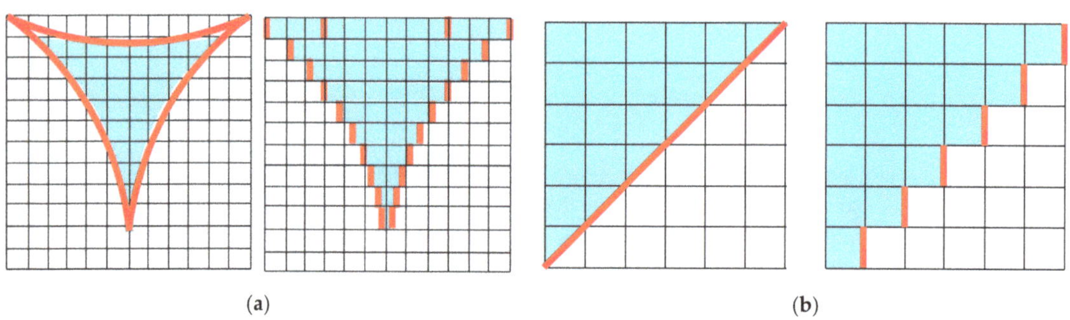

Figure 10. Cross-sectional view of the true and divided shapes of foam frame: (**a**) inside part of three circumscribed circles and (**b**) inclined flat surface.

3.4. Propagation Constant and Characteristic Impedance

The propagation constant and characteristic impedance are derived by considering the attenuation of sound waves in the voids for which two-plane approximation was performed. Tjideman [25] and Stinson [26] studied circular tubes, Stinson [27] studied equilateral triangle tubes, and Beltman [28] studied rectangular tubes. Moreover, Allard [29] considered the degree of tortuosity in his analysis. The methods of Stinson [27] and Allard [29] were applied in this study.

The Cartesian coordinate system is considered (Figure 11), and a three-dimensional analysis is performed using the Navier–Stokes equations, gas equation of state, continuity equation, energy equation, and dissipative function representing heat transfer to derive the effective density ρ_s and compression ratio C_s, as shown in Equations (2) and (3), respectively:

$$\rho_s = \rho_0 \left[1 - \frac{tanh(\sqrt{j}\lambda_s)}{\sqrt{j}\lambda_s} \right]^{-1}, \ \lambda_s = \frac{b_n}{2}\sqrt{\frac{\omega\rho_0}{\eta}} \tag{2}$$

$$C_s = \left(\frac{1}{\kappa P_0}\right)\left\{1 + (\kappa - 1)\left[\frac{tanh(\sqrt{jN_{pr}}\lambda_s)}{\sqrt{jN_{pr}}\lambda_s}\right]\right\} \tag{3}$$

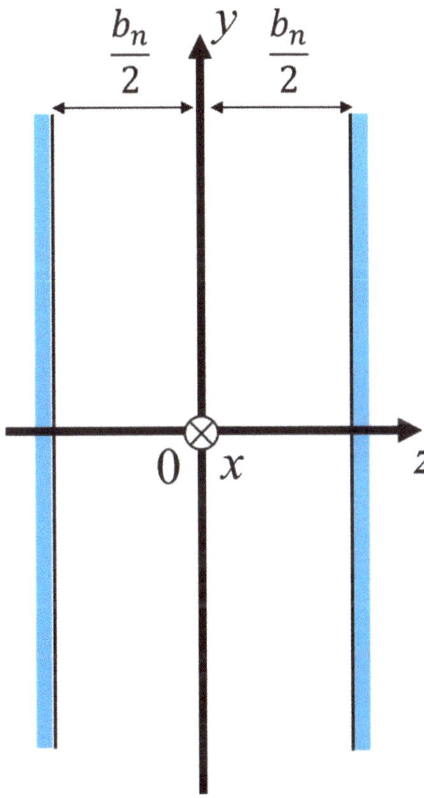

Figure 11. Cartesian coordinate system for parallel space between two planes.

These values were derived from the work of Stinson [26] and Allard [29], where ρ_0 denotes the air density, λ_s denotes the mediator variable, b_n denotes the thickness of the air clearance between the two planes, ω denotes the angular frequency, η denotes the air viscosity, κ denotes the specific heat ratio of air, P_0 denotes the atmospheric pressure, and N_{pr} denotes the Prandtl number.

Using the effective density ρ_s and compressibility C_s outlined above, the propagation constant γ and characteristic impedance Z_c are expressed by the following equations, based on the definitions provided by Stinson [26,27] and others:

$$\gamma = j\omega\sqrt{\rho_s C_s} \tag{4}$$

$$Z_c = \sqrt{\frac{\rho_s}{C_s}} \tag{5}$$

By the effective density multiplied by the tortuosity, the propagation constant and characteristic impedance considering the tortuosity can be obtained [29]. Therefore, the propagation constant and characteristic impedance when considering the tortuosity α_∞ can be expressed as follows:

$$\gamma = j\omega\sqrt{\alpha_\infty \rho_s C_s} \tag{6}$$

$$Z_c = \sqrt{\frac{\alpha_\infty \rho_s}{C_s}} \tag{7}$$

3.5. Transfer Matrices and Their Connections

The clearance between the two planes was analyzed using the transfer matrix method for sound pressure and volume velocity based on a one-dimensional wave equation. Figure 12 shows a schematic of one element in the x-direction for the clearance between the two planes shown in Figure 11.

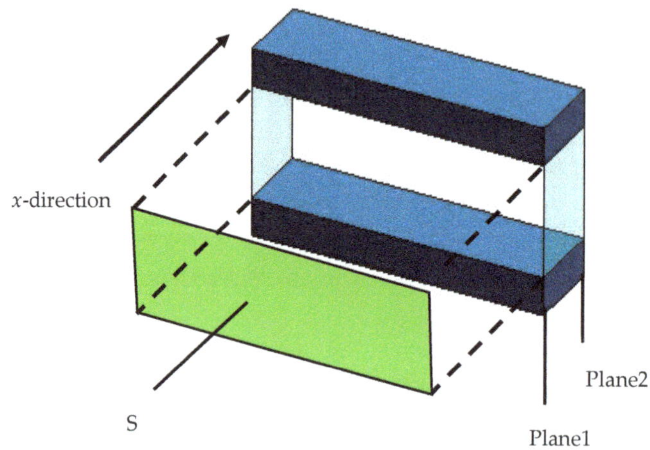

Figure 12. Sound incident area, incident plane, and transmission plane of approximated clearance between the two planes.

Using the cross-sectional area S of the clearance, pitch d per layer, characteristic impedance Z_c, and propagation constant γ, the transfer matrix T and four-terminal constants A~D of the acoustic tube element can be expressed as follows:

$$T = \begin{bmatrix} cosh(\gamma d) & \frac{Z_c}{S} sinh(\gamma d) \\ \frac{S}{Z_c} sinh(\gamma d) & cosh(\gamma d) \end{bmatrix} = \begin{bmatrix} A & B \\ C & D \end{bmatrix}. \quad (8)$$

Plane 1 is the sound wave incidence plane, and Plane 2 is the sound wave transmission plane; if the sound pressure and particle velocity in Plane 1 are p_1 and u_1, respectively, and the sound pressure and particle velocity in Plane 2 are p_2 and u_2, respectively, the transfer matrix is expressed as follows:

$$\begin{bmatrix} p_1 \\ Su_1 \end{bmatrix} = \begin{bmatrix} A & B \\ C & D \end{bmatrix} \begin{bmatrix} p_2 \\ Su_2 \end{bmatrix}. \quad (9)$$

By applying Equation (9) to the clearance between the two planes derived by image processing, the transfer matrix for each segmented element is derived.

Based on the equivalent circuit shown in Figure 13, the transfer matrix T_{all} for the entire sample is derived by connecting the transfer matrices of each successive segment in the x-axis direction in the cascade. The number of transfer matrices connected in the cascade is equal to the number of images, i.e., the image pitch multiplied by the number of images gives the height of each sample.

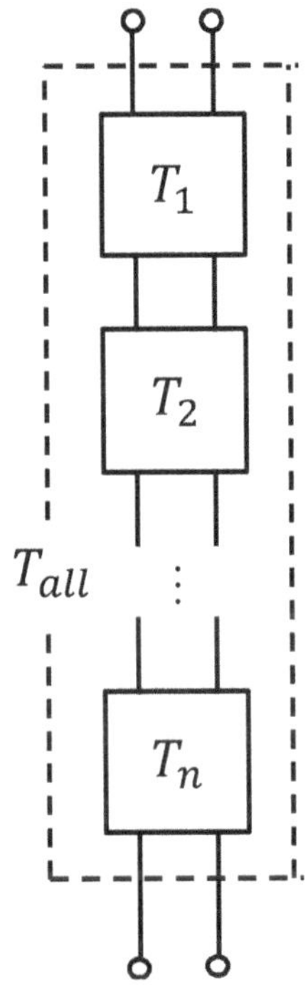

Figure 13. Cascade connecting T_{all}.

3.6. Derivation of Normal Incident Sound Absorption Coefficient

The sound absorption coefficient is derived from the transfer matrix T_{all} obtained in Section 3.5. Applying the acoustic tube of Figure 11 to the actual sample, Plane 2 is a rigid wall. Therefore, because the particle velocity $u_2 = 0$, Equation (9) can be transformed as in Equation (10) to obtain Equation (11).

$$\begin{bmatrix} p_1 \\ Su_1 \end{bmatrix} = \begin{bmatrix} A & B \\ C & D \end{bmatrix} \begin{bmatrix} p_2 \\ 0 \end{bmatrix} \tag{10}$$

$$\begin{bmatrix} p_1 \\ Su_1 \end{bmatrix} = \begin{bmatrix} Ap_2 \\ Cp_2 \end{bmatrix} \tag{11}$$

According to Equation (10), the specific acoustic impedance Z_1 looking into the interior of the sample from Plane 1 can be expressed as follows:

$$Z_1 = \frac{p_1}{u_1} = \frac{p_1}{Su_1}S = \frac{A}{C}S. \tag{12}$$

The relationship between the specific acoustic impedance Z_0 and reflectance R is expressed as follows:

$$R = \frac{Z_1 - \rho_0 c_0}{Z_1 + \rho_0 c_0} \tag{13}$$

The following relationship between the sound absorption coefficient and reflectance and Equation (13) gives the theoretical value of a sample's normal incident sound absorption coefficient α:

$$\alpha = 1 - |R|^2 \tag{14}$$

4. Comparison of Calculated and Measured Results

4.1. Theoretical Values Obtained by CT Images with Varying Binarization Thresholds

The theoretical values obtained based on the CT images with varying binarization thresholds (Figure 8) are compared with the measured values. The sound absorption coefficient based on CT images with varying binarization thresholds is shown in Figure 14, together with the measured values.

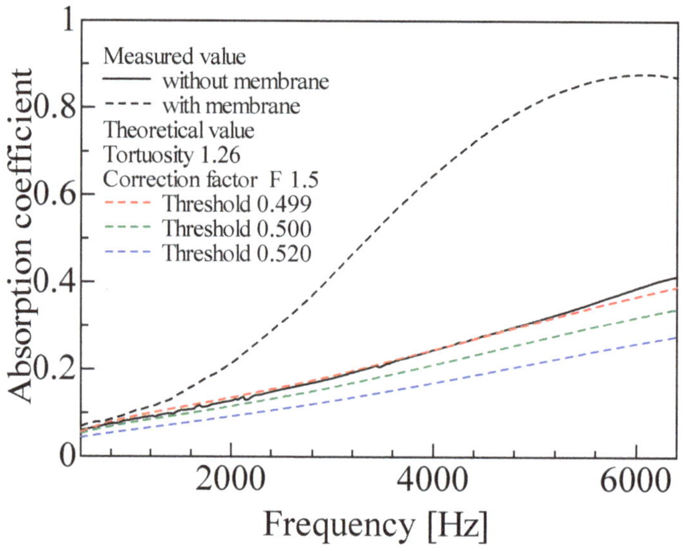

Figure 14. Comparison of experimental and theoretical results using images scanned by MCT225 Metrology CT: theoretical values due to changes in threshold.

For reference, Figure 14 also shows the experimental results for a foam sound absorber with the membrane present. The sound absorption coefficient is higher in all frequency ranges when the membrane is present than after the membrane removal.

Figure 14 shows changes in the sound absorption coefficient due to changes in the threshold value. As shown in Figure 8, this is thought to be due to differences in the amount of residual noise and recognition of the form skeleton cross-section caused by changes in the threshold value, resulting in changes in the sound absorption coefficient. As the threshold is decreased, the fin portion can be extracted, but the noise increases in the entire binarized image. Meanwhile, if the noise reduction effect is strengthened by increasing the filter or threshold value, the fin portion is also removed as noise. This problem is attributable to the lack of image resolution for the smallest foam material details, which makes it difficult to distinguish between the noise and fin portions.

4.2. Sensitivity Analysis of the Correction Factor for Theoretical Values Calculated from CT Images

For the reasons given in Section 3.3, a sensitivity analysis was performed on the correction factor F to approximate the true surface area of the skeleton. That is, the threshold is fixed at a representative value, and the theoretical values are compared with the measured values while varying the correction factor F.

Figure 15 shows the sound absorption coefficient for varying the correction factor F in CT images. For reference, the theoretical values are shown when both the tortuosity and correction factor F are not considered. In this case, the lowest sound absorption coefficient is shown in Figure 15.

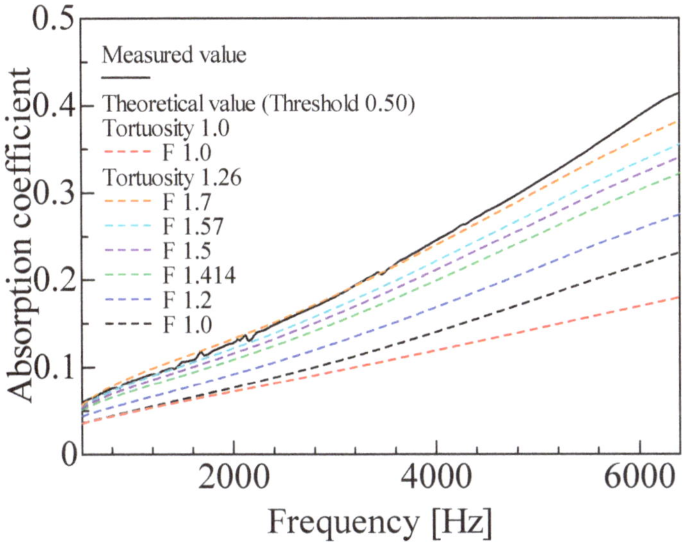

Figure 15. Comparison of experimental and theoretical results using images scanned by MCT225 Metrology CT: theoretical values due to changes in the correction factor.

The sound absorption coefficient increases with the correction factor *F* because the surface area reflected in the analysis increases with the correction factor *F*.

According to Section 3.3 and previous studies [19], a correction factor *F* of approximately 1.4–1.6 is considered suitable. However, in Figure 14, the best agreement between the theoretical and experimental values is found when the correction factor *F* is approximately 1.7. In the foam materials with removed membranes, it is considered that the membranes are not completely removed but remain partially in place. Consequently, if the remaining membranes and other details below the CT imaging resolution are not fully captured, as in the CT images in this study, the calculated sound absorption coefficient is expected to be smaller than the experimental value. A further discussion of the differences between experimental and theoretical values is presented in Section 4.3.

4.3. Discussion of Differences Between Measured and Theoretical Values

Based on the above results, the differences between the measured and theoretical values are discussed.

The first cause is the image processing complexity: during the preparation stage of the CT scan image binarization, an attempt was made to apply Otsu's binarization [30], a typical method for determining threshold binarization values, but the fin portion was excessively removed. Therefore, a threshold was determined to an appropriate value while visually observing an image, and the image was processed to improve the recognition of the fin and other parts. As shown above, the recognition of fin portions is sensitive to the threshold value, and whether the fin surface area is close to the true value or not influences the difference from the measured value.

The second cause is that the samples, used for the CT scanning and sound absorption coefficient measurement, were cut from the same piece of material; however, they are strictly portions of different locations.

The third cause is the lack of detail extraction. In Figure 14, the theoretical value for the 0.499 threshold best agrees with the experimental value. However, examining the image after binarization, the 0.500 threshold in Figure 8b is the most appropriate due to the combination of noise removal and detail extraction. The shortfall of the theoretical value from the binarized image with the 0.500 threshold compared with the experimental value is attributable to the lack of detail extraction. Here, the fact that the 0.499 threshold case is closest to the experimental value is assumed to be because the excess residual noise compensates for the lack of detail extraction.

5. Conclusions

The normal incident sound absorption coefficient was theoretically derived using CT images of foam sound absorbers with the membranes removed. The circumference of the foam sound absorber framework and the cross-sectional area of voids were determined by CT image processing, and the propagation constant and characteristic impedance were derived. Tortuosity, which indicates the complexity of the sound wave propagation path, was measured and introduced into the theoretical analysis. The measured and theoretical values of the normal incident sound absorption coefficient were compared and evaluated.

Theoretical analysis was performed based on CT images of foam sound absorbers for which geometrical modeling is difficult due to the lack of strict regularity in the structure, and the sound absorption coefficient was theoretically derived. By varying the correction factor for the skeleton surface area, the theoretical value of the sound-absorbing foam without a membrane was closer to the measured value. Therefore, the theoretical estimation of the sound absorption coefficient using CT images can be useful.

Author Contributions: Conceptualization, S.S.; software, K.M. and Y.H.; formal analysis, T.S. and K.T.; data curation, T.S., K.T., K.M. and Y.H.; supervision, S.S.; project administration, S.S. All authors have read and agreed to the published version of the manuscript.

Funding: This study was supported by FUKOKU Co., Ltd.

Institutional Review Board Statement: Not applicable.

Informed Consent Statement: Not applicable.

Data Availability Statement: The original contributions presented in this study are included in the article. Further inquiries can be directed to the corresponding author.

Acknowledgments: This study was conducted in collaboration with FUKOKU Co., Ltd.

Conflicts of Interest: Author Takamasa Satoh was employed by the company Fukoku Co., Ltd. The remaining authors declare that the research was conducted in the absence of any commercial or financial relationships that could be construed as a potential conflict of interest.

References

1. Fischer, J.; Behrendt, M.; Lieske, D.; Albers, A. Measurement and analysis of the interior noise and the transfer path of acoustic phenomena into the driver cabin of a battery electric vehicle. In Proceedings of the INTER-NOISE 2014, Melbourne, Australia, 16–19 November 2014. Paper No. 110.
2. Kim, J.G.; Kim, S.K.; Kim, J.H. Sound absorption behavior of flexible polyurethane foams with distinct cellular structures. *Mater. Des.* **2016**, *89*, 448–454. [CrossRef] [PubMed]
3. Yamashita, T.; Suzuki, K.; Nishino, S.; Tomota, Y. Relationship between Sound Absorption Property and Microscopic Structure Determined by X-ray Computed Tomography in Urethane Foam Used as Sound Absorption Material for Automobiles. *Mater. Trans.* **2008**, *49*, 345–351. [CrossRef]
4. Bérengier, M.C.; Stinson, M.R.; Daigle, G.A.; Hamet, J.F. Porous Road Pavements: Acoustical Characterization and Propagation Effects. *J. Acoust. Soc. Am.* **1997**, *101*, 155–162. [CrossRef]
5. Licitra, G.; Cerchiai, M.; Teti, L.; Ascari, E.; Fredianelli, L. Durability and Variability of the Acoustical Performance of Rubberized Road Surfaces. *Appl. Acoust.* **2015**, *94*, 20–29. [CrossRef]
6. Ciaburro, G. Recycled Materials for Sound Absorbing Application. *Mater. Sci. Forum* **2021**, *1034*, 169–175. [CrossRef]
7. Mitrevska, M.J.; Mickovski, V.; Samardzioska, T.; Iannace, G. Experimental and Numerical Investigation of Sound Absorption Characteristics of Rebonded Polyurethane Foam. *Appl. Sci.* **2022**, *12*, 12936. [CrossRef]

8. Shinde, N.G.; Patel, D.M. A Short Review on Automobile Dashboard Materials. *IOP Conf. Ser. Mater. Sci. Eng.* **2020**, *810*, 012033. [CrossRef]
9. Patten, W.N.; Sha, S.; Mo, C. A Vibrational Model of Open-Celled Polyurethane Foam Automotive Seat Cushions. *J. Sound Vib.* **1988**, *217*, 145–161. [CrossRef]
10. del Rey, R.; Alba, J.; Arenas, J.P.; Sanchis, V.J. An empirical modelling of porous sound absorbing materials made of recycled foam. *Appl. Acoust.* **2012**, *73*, 604–609. [CrossRef]
11. Iwase, T. Sound absorption properties of porous materials and measurement technique for them. *J. Acoust. Soc. Jpn.* **1997**, *54*, 66–71. (In Japanese)
12. Trinh, V.H.; Langlois, V.; Guilleminot, J.; Perrot, C.; Khidas, Y.; Pitois, O. Tuning membrane content of sound absorbing cellular foams: Fabrication, experimental evidence and multiscale numerical simulations. *Mater. Des.* **2019**, *162*, 345–361. [CrossRef]
13. Zhang, C.; Li, J.; Hu, Z.; Zhu, F.; Huang, Y. Correlation between the acoustic and porous cell morphology of polyurethane foam: Effect of interconnected porosity. *Mater. Des.* **2012**, *41*, 319–325. [CrossRef]
14. Park, J.H.; Minn, K.S.; Lee, H.R.; Yang, S.H.; Yu, C.B.; Pak, S.Y.; Oh, C.S.; Song, Y.S.; Kang, Y.J.; Youn, J.R. Cell openness manipulation of lowdensity polyurethane foam for efficient sound absorption. *J. Sound Vib.* **2017**, *406*, 224–236. [CrossRef]
15. Gaulon, C.; Pierre, J.; Derec, C.; Jaouen, L.; Becot, F.-X.; Chevillotte, F.; Elias, F.; Drenckhan, W.; Leroy, V. Acoustic absorption of solid foams with thin membranes. *Appl. Phys. Lett.* **2018**, *112*, 261904. [CrossRef]
16. Satoh, T.; Sakamoto, S.; Iizuka, K.; Tasaki, K.; Tanabe, K. Mathematical Modeling and Experiments on the Acoustic Properties of a Frame Structure with a Membrane. *Noise Control Eng. J.* **2024**, *72*, 297–315. [CrossRef]
17. Sakamoto, S.; Tanabe, K.; Maruyama, K.; Sato, T. Estimation and experiment of the sound absorption coefficient of foam sound-absorbing material (Estimation of sound absorption coefficient using CT scan images). In Proceedings of the 17th International Conference on Motion and Vibration Control and the 20th Asia-Pacific Vibration Conference (MoViC & APVC2024), Tokyo, Japan, 5–8 August 2024. 5p, Paper No. U00126.
18. Sakamoto, S.; Suzuki, K.; Toda, K.; Seino, S. Estimation of the Acoustic Properties of the Random Packing Structures of Granular Materials: Estimation of the Sound Absorption Coefficient Based on Micro-CT Scan Data. *Materials* **2023**, *16*, 337. [CrossRef]
19. Sakamoto, S.; Toda, K.; Seino, S.; Hoshiyama, K. Theoretical and Experimental Analyses on the Sound Absorption Coefficient of Rice and Buckwheat Husks Based on Micro-CT Scan Data. *Materials* **2023**, *16*, 5671. [CrossRef]
20. Sakamoto, S.; Seino, S.; Hoshiyama, K.; Kaneko, K.; Kojima, Y. Estimation and Experiment on Acoustic Properties of Rice Straw (Estimation of Sound Absorption Coefficient Based on CT Image). In Proceedings of the 53rd International Congress and Exposition on Noise Control Engineering (INTER-NOISE 2024), Nantes, France, 25–29 August 2024. 6p, Paper No. 4064.
21. Satoh, T.; Sakamoto, S.; Isobe, T.; Iizuka, K.; Tasaki, K. Mathematical Model for Estimating the Sound Absorption Coefficient in Grid Network Structures. *Materials* **2023**, *16*, 1124. [CrossRef]
22. *ISO 10534-2*; Acoustics—Determination of Acoustic Properties in Impedance Tubes. Part 2: Two-Microphone Technique for Normal Sound Absorption Coefficient and Normal Surface Impedance. ISO: Geneva, Switzerland, 2023.
23. Sakamoto, S.; Suzuki, K.; Toda, K.; Seino, S. Mathematical Models and Experiments on the Acoustic Properties of Granular Packing Structures (Measurement of tortuosity in hexagonal close-packed and face-centered cubic lattices). *Materials* **2022**, *15*, 7393. [CrossRef]
24. Canny, J. A Computational Approach to Edge Detection. *IEEE Trans. Pattern Anal. Mach. Intell.* **1986**, *8*, 679–698. [CrossRef]
25. Tijdeman, H. On the propagation of sound waves in cylindrical tubes. *J. Sound Vib.* **1975**, *39*, 1–33. [CrossRef]
26. Stinson, M.R. The propagation of plane sound waves in narrow and wide circular tubes and generalization to uniform tubes of arbitrary cross-sectional shape. *J. Acoust. Soc. Am.* **1991**, *89*, 550–558. [CrossRef]
27. Stinson, M.R.; Champou, Y. Propagation of sound and the assignment of shape factors in model porous materials having simple pore geometries. *J. Acoust. Soc. Am.* **1992**, *91*, 685–695. [CrossRef]
28. Beltman, W.M.; van der Hoogt, P.J.M.; Spiering, R.M.E.J.; Tijdeman, H. Implementation and Experimental Validation of a New Viscothermal Acoustic Finite Element for Acousto-Elastic Problems. *J. Sound Vib.* **1998**, *216*, 159–185. [CrossRef]
29. Allard, J.F. Propagation of sound in Porous Media Modeling Sound Absorbing Materials. *J. Acoust. Soc. Am.* **1994**, *95*, 2785. [CrossRef]
30. Otsu, N. A Threshold Selection Method from Gray-Level Histograms. *IEEE Trans. Syst. Man Cybern.* **1979**, *9*, 62–66. [CrossRef]

materials

MDPI

Article

Acoustic Performance of Tufted Carpets Coupled with Underlayment Produced from Tannery Wool Waste

Jan Broda *, Katarzyna Kobiela-Mendrek, Marcin Baczek and Monika Rom

Faculty of Materials, Civil and Environmental Engineering, University of Bielsko-Biala, Willowa 2, 43-309 Bielsko-Biala, Poland; kkobiela@ubb.edu.pl (K.K.-M.); mbaczek@ubb.edu.pl (M.B.); mrom@ubb.edu.pl (M.R.)

* Correspondence: jbroda@ubb.edu.pl; Tel.: +48-33-8279117

Abstract: Sheep wool is a precious, renewable raw material that is nowadays disregarded and wasted. To better use local sources of wool, it was used to manufacture tufted carpets. The coarse wool of mountain sheep was used to form a carpet pile layer, while the waste wool from the tannery industry was applied to form carpet underlayment. During investigations, the acoustic performance of the carpets was assessed. The carpets' sound absorption coefficients and transmission loss were determined using the impedance tube. It was revealed that the adding of underlayment improves the carpet's sound absorption only at medium sound wave frequencies. The underlayment significantly increases transmission loss in the whole frequency range. The acoustic performance of the carpets with the wool underlayment is similar to the acoustic characteristics of the carpets with an underlayment made from polyester. It was concluded that wool nonwovens can be used as an effective, eco-friendly, sound-absorbing carpet underlayment, which can improve wool utilisation and contribute to the reduction in environmental pollution caused by plastic residues.

Keywords: wool; carpet; underlayment; sound absorption; transmission loss

1. Introduction

Industrial activities, road, rail, and air traffic, construction works, and many other human activities generate sounds, which are the source of environmental noise. Excessive noise has become a severe and pervasive pollutant that harmfully influences physical and mental health [1]. In addition to external noise, the inner noise in closed rooms has become a severe problem. Exposure to domestic noise negatively affects human well-being, impairs productivity, generates higher stress, and contributes to somatic complaints. Given the adverse effects of noise, the urgent need to reduce noise levels in homes, classrooms, open-plan offices, workplaces, and public facilities is highly significant and desirable today.

For many years, various textiles used as window curtains, tapestries, upholstery, wall and ceiling panels, screens, rugs, and carpets have been used to reduce noise and improve the acoustic comfort of interiors [2–5]. Among these products, carpets are the most versatile, controlling indoor noise in various ways. First, carpets lower the level of noise by absorbing airborne sound. Second, they reduce the generation of floor impact sounds produced by footfalls, furniture movement, and objects dropped onto the floor. Third, carpets minimise the noise transmission through floors to adjoining rooms in multi-storied buildings.

Pile carpets, in particular, demonstrate a significant potential for noise reduction. These carpets are made of two layers that perform different functions and have different structures. The top layer, subjected to foot traffic, comprises overhanging yarn segments arranged in a compact system formed from loops or upright tufts. Instead of reflecting sound waves

https://doi.org/10.3390/ma18020315

into the room, this layer allows their penetration into the piles. The penetrating waves cause the vibration of air molecules between both the individual fibres and pile tufts. The oscillating air molecules rub against irregular interstices and lose their energy due to friction. Eventually, the sound wave's energy is transformed into thermal and viscous heat, which is finally dissipated. The energy dissipation in the pile layer is most effective at high sound frequencies [6].

The sound absorption in the pile layer depends on the pile type (loop or cut) and its height and density. Due to the open fuzzy structure, carpets with cut piles have higher sound absorption capacity than carpets with loop piles. Increasing pile height and density increases the interphase contact surface between piles and vibrating air molecules. Then, the energy dissipated by friction is higher and the resulting sound absorption increases [7–9]. The influence of other parameters characterising the pile layer, such as fibre type, yarn parameters, or knot type in knotted-pile carpets, is less distinguishable and difficult to estimate [10].

The second layer, the carpet base, provides carpets with appropriate dimensions and dimensional stability. The base consists of a woven fabric for woven and knotted-pile carpets. For carpets manufactured by tufting, the base is a multi-layered composite built from the primary backing, a layer of adhesive compound, and secondary backing fabric. The primary backing, formed from fabric with a mesh structure, is the carpet's foundation, which locks the tufted yarns inserted by a needle. The secondary backing covers the loops of the tufted yarns, protects the tufts from being pulled out, and enhances the carpet's final dimensional stability, strength, and stretch resistance. The base's structure and the finishing type affect the absorption and transmission of sounds and significantly contribute to the carpet's overall acoustic properties. The acoustic performance of the base, like that of all other weave fabrics, depends on the weave structure and other structural parameters, including cover factor and yarn density [11]. The base, which possesses distinct geometry and is formed from fibres other than the pile layer, has different resonant frequencies in which it absorbs sound. Consequently, the maximal sound absorption falls in other ranges, potentially improving the sound absorption coefficients at low and medium sound frequencies. The sound absorption at lower frequencies is enhanced for a layered base formed from two woven fabrics separated by an air gap [12].

In addition to the two essential layers, top and base, certain carpets are coupled with additional underlayment formed from nonwovens, felt or foam. The underlayment provides a barrier between the carpet and the floor, cushions footfalls, and extends carpet life. Simultaneously, it secures carpets with better thermal insulation and significantly reduces airborne and structure-borne noise. Adding underlay increases the sound absorption, principally in the mid-frequency range [13]. Soft underlay effectively lessens the force of the impact of footsteps and objects dropped onto the floor. The underlay layer reduces impact noise and its transmission through the floor, significantly contributing to the room's soundproofing.

Pile carpets are manufactured from various natural and synthetic fibres. Although synthetic products dominate today's market, the most valued and luxurious carpets are made from sheep wool [14]. These carpets possess high aesthetic values and are distinguished by high durability, flame resistance, and other beneficial functional properties [15–18].

A coarser type of wool, the so-called carpet wool, is applied to produce carpets [19,20]. The thickness of this wool exceeds 35 μm, making it, contrary to Merino wool and other fine and soft wool types, useless for producing apparel textiles. The appropriate wool, which meets manufacturing requirements and ensures superior carpet appearance and performance, is sourced worldwide from different sheep breeds. Often, suitable wool is obtained from local breeds, which deliver wool that is nowadays undervalued by the market and frequently treated as a waste of sheep husbandry [21,22].

In ordinary practice, wool, which is a sustainable, renewable, and biodegradable raw material, is used to produce rug yarns designed to construct only the pile layer. The remaining carpet layers, base, and underlayment are made from other natural or synthetic materials. Due to their excellent mechanical properties, chemical and biological resistance, and low price, fabrics made from synthetic fibres, mainly polyester or polypropylene, are often applied as the base. As the underlayment, polyurethane or polyethylene foams or polyester nonwovens are used primarily. Applying synthetic materials generates non-biodegradable waste and contributes to the pollution of the environment with plastic residues. For the sake of environmental concern, reducing the amount of plastic and replacing it with biodegradable materials is highly desirable [23].

To reduce the application of plastic and eliminate largely unexploited wool waste, investigations into the use of wool to produce carpets were performed. In previous investigations, coarse wool from Polish mountain sheep was applied to produce a pile layer of tufted carpets. The rug yarns were manufactured, and carpets with cut and loop piles were obtained. The investigations revealed that the acoustic properties are determined by the pile type and height and the density of the pile layer [24,25]. As a continuation of these investigations, studies of the acoustic performance of carpets coupled with additional underlayment were carried out. The underlayment was made from tannery waste wool. This wool is obtained by dehairing the sheep hide as a by-product of the tanning processes [26]. The wool is too short to be used in textile processes and, in many countries, is treated as troublesome and useless waste. A few papers report attempts to use this waste to produce thermal insulating composites or fertilisers [27,28].

Needle-punched nonwovens from waste tannery wool were obtained and applied as carpet underlayment. The effectiveness of the wool underlayment in sound absorption and transmission loss of the carpet was compared with the performance of products coupled with polyester nonwoven fabrics.

2. Materials and Methods

2.1. Materials

The investigations were carried out on the carpets, consisting of three layers: pile layer, base, and underlayment.

The pile layer and the carpet base were obtained similarly as in studies reported in previous papers [24,25]. The piles were made from the coarse wool of Polish mountain sheep raised in the Silesian Beskid Mountains in southern Poland. The wool was collected during shearing and separated from heavily contaminated wool and wool of worse quality, designed to be used as mulch or fertilisers. The wool is a mixture of thinner, shorter, and more delicate fibres with diameters varying between 35 and 40 μm, thicker and longer guarded hairs containing a significant amount, ca. 10%, of medullated fibres, and kemp [29,30].

The wool was used to manufacture semi-worsted ring-spun three-plied yarn with a linear density of 1240 tex. The yarn was manufactured in the Selbu Spinneri (Klæbu, Norway), using a small-scale technological line of carding, drafting, and spinning machines set by Belfast Mini Mills (Prince Edward Island, Canada) and Ramella (Biella, Italy).

The yarn was inserted into the carpet base using a manual pneumatic tufting gun in industrial conditions in the Sztuka Beskidzka (Czechowice-Dziedzice, Poland). The yarns were punched with a tubular needle into a pre-stretched primary backing (PB) mounted on a vertical frame. Then, a secondary backing (SB) was bonded with 520 g/m^2 latex adhesive, commonly used for finishing tufted carpets. A plain polyester weave cloth with a square set of 70 dm^{-1} and yarn linear density of 180 tex was applied as the primary backing. As the secondary backing, a leno woven fabric with a mesh structure consisting of polypropylene warp with a linear density of 33 tex and set equal to 64 dm^{-1} and polyacrylonitrile weft with a linear density of 125 tex and set equal to 40 dm^{-1} was used.

Two sets of carpets with cut and loop piles with a height of 12 mm were manufactured. The tuft density for carpets with cut piles was about 42·10 dm^{-2}. For loop piles, the density was lower, approximately 40·10 dm^{-2}.

Three nonwovens of different masses per unit area made from waste wool were applied as carpet underlayment. The wool was a waste product from the tanning industry and a mixture of fibres of various lengths and thicknesses from different Norwegian sheep breeds. The nonwovens were manufactured using a needle-punching technique by interlocking the carded web with the one-side needling machine with a punching density of 28 cm^{-2}. The nonwovens were produced in industrial conditions, using that technological line (Jiangsu XST Machinery Technology, Wuxi, China).

Additionally, the commercial needle-punched nonwoven made from polyester, commonly used as carpet underlayment, was applied as a control.

Finally, the multi-layered sandwich structures were produced by bonding the underlayment with latex adhesive (Figure 1).

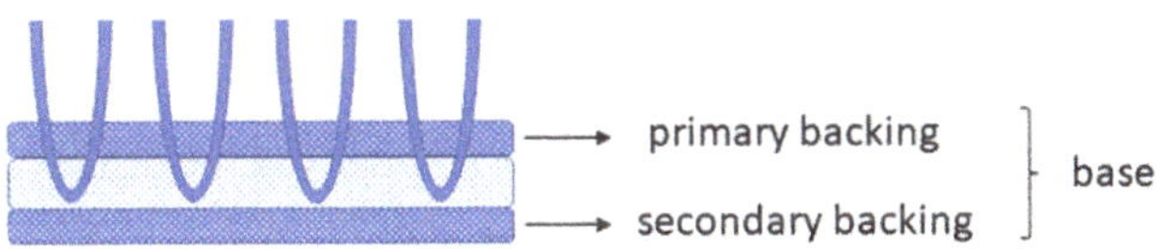

(a)

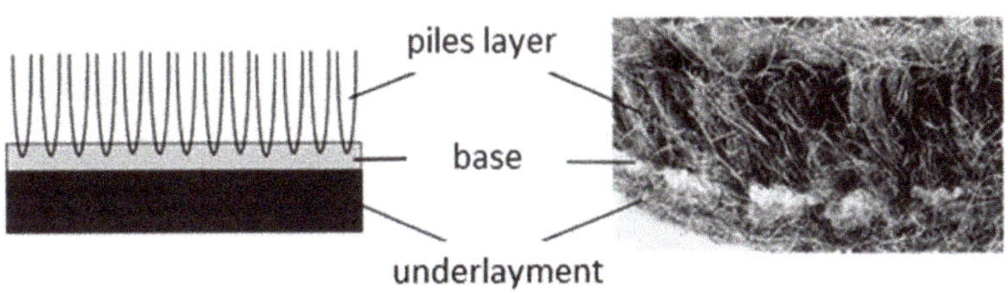

(b)

Figure 1. The carpet construction: (**a**) the base layer; (**b**) the composite multi-layered structure.

2.2. Methods

Surface density and thickness were measured to characterise nonwovens used as carpet underlayment. The measurements were carried out according to Polish standards PN-EN ISO 9863-1:2007 and PN-EN ISO 9864:2007. The thickness was measured under a 2 kPa load using a Tilmet 73 thickness gauge (Lodz, Poland). Each parameter was determined as the mean of 10 measurements.

The morphology of the nonwovens was studied using Scanning Electron Microscopy. Before the observations, the samples were covered with a thin layer of gold in a Leica EM ACE 200 low-vacuum coater (Wetzlar, Germany). A high-resolution Phenom ProX SEM microscope (PhenomWorld, Eindhoven, The Netherlands), operated in a backscattered electron mode, was used for the observations.

The acoustic properties of the pile carpets and nonwovens used as carpet underlayment were characterised by the normal incidence sound absorption coefficient (SAC) and normal incidence sound transmission loss (STL). The parameters were measured according to the relevant standards, PN-EN ISO 10534-2 and ASTM E2611, using the BSWA impedance tube system SW422+SW477 (BSWA Technology Co., Ltd., Beijing, China). The measuring stand included two impedance tubes with inner diameters of 30 mm (SW422) and 100 mm (SW477), MI 19 microphones (¼ " ICP), a MC3242 analyser, and a PA50 amplifier. A scheme of system configurations for the measurement of the sound absorption coefficient (SAC) and sound transmission loss (STL) is presented in Figure 2.

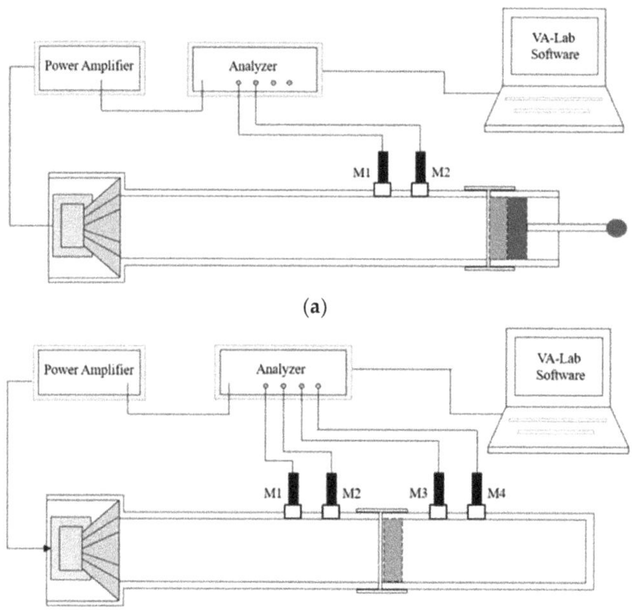

(b)

Figure 2. Scheme of experimental setup used for acoustic measurements of (**a**) sound absorption coefficient measurement, (**b**) sound transmission loss measurement.

The measurements were carried out at the incident wave's one-third-octave band frequencies within the 80–5000 Hz range. Two measurements were performed for each sample, and mean values were calculated. Based on the measurement results, the sound absorption average (SAA) was calculated as the arithmetic mean of the sound absorption coefficients for the twelve one-third-octave bands from 200 to 2500 Hz.

3. Results

3.1. Acoustic Properties of Nonwovens Used as Carpet Underlayment

The nonwovens used as carpet underlayment are characterised by the parameters presented in Table 1. The mass per unit area of the wool nonwovens was from 200 to 400 g/m^2. The thickness of the nonwoven with the lowest mass equalled only 3.6 mm. For the nonwovens with higher mass, the thickness was 1.5 mm higher. The characteristics of the woollen nonwoven with the lowest mass corresponds well with the parameters of the polyester nonwoven, which is ordinarily used as underlayment in commercial products.

Figure 3 presents the SEM images of the nonwovens. As a waste of the tanning industry, the wool is a mixture of thinner and coarser fibres with diameters between 18 and 50 µm. The thinner fibres are flexible and easily bent, forming an entangled structure. The coarser and stiffer fibres are incorporated into a tangled network as straight fibres without bends. The surface of the wool fibres is covered with characteristic scales. The scales facilitate the mutual interlocking of the fibres, and the creation of a compact felt structure.

Table 1. Specifications of nonwovens used as carpet underlayment.

Symbol	Material	Mass per Unit Area, g·m^{-2}	Thickness, mm
W1	wool	200 ± 25	3.6 ± 0.3
W2	wool	300 ± 30	4.9 ± 0.4
W3	wool	400 ± 40	5.2 ± 0.5
PET	polyester	200 ± 20	4.1 ± 0.3

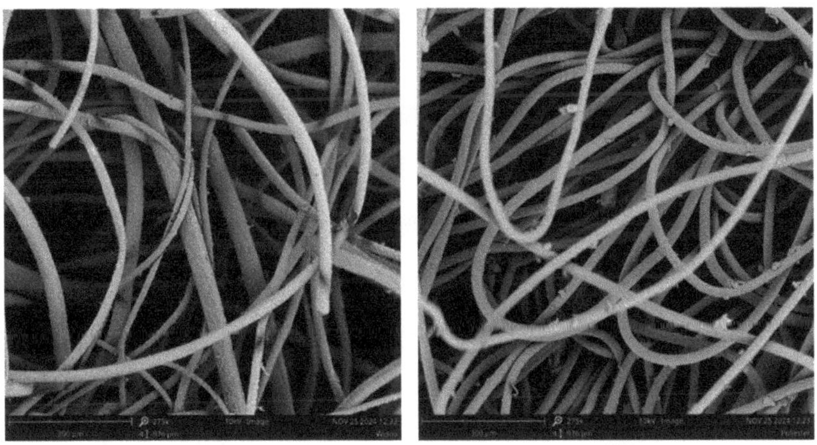

(a) (b)

Figure 3. SEM images of nonwovens made from (**a**) wool and (**b**) polyester.

Unlike wool, polyester fibres are uniform, with a diameter of 25 µm. The fibres are sufficiently flexible to form a tangled web. The fibres possess a smooth surface without scales, grooves, or other irregularities affecting the surface roughness. Due to their smooth surface, the fibres can slide easily against each other, so the particular fibres do not stick together and are well separated.

The absorption coefficients measured for the nonwovens change with the sound wave frequency from 0.03 to 0.2 (Figure 4). At low and medium frequencies up to 1600 Hz, the sound absorption coefficient is low and does not exceed 0.1. The SAC slightly fluctuates in this range and is almost the same for all the investigated fabrics. At higher frequencies above 1600 Hz, with increasing frequency, the SAC increases gradually, reaching, for all fabrics, a maximum for the highest frequency of 5000 Hz. For the two wool nonwovens with a higher mass (300 and 400 g/m^2), the maximal value of the SAC equalled 0.2 is approximately the same as the coefficient measured for the nonwoven polyester. The maximal SAC is ca. 25 % lower for the wool nonwoven with the lowest mass (200 g/m^2).

The sound transmission loss (STL) for all nonwovens in the whole frequency range takes very low values (Figure 5).

The STL remains below 1 dB at low and medium frequencies. The STL takes on slightly higher values at the higher frequencies, reaching a maximum value at the end of the measurement range. The maximal value for the wool nonwoven with the lowest mass (200 g/m^2) is slightly lower than that of polyester. For the two wool nonwovens with higher masses (300 and 400 g/m^2), the STL is almost 1 dB higher.

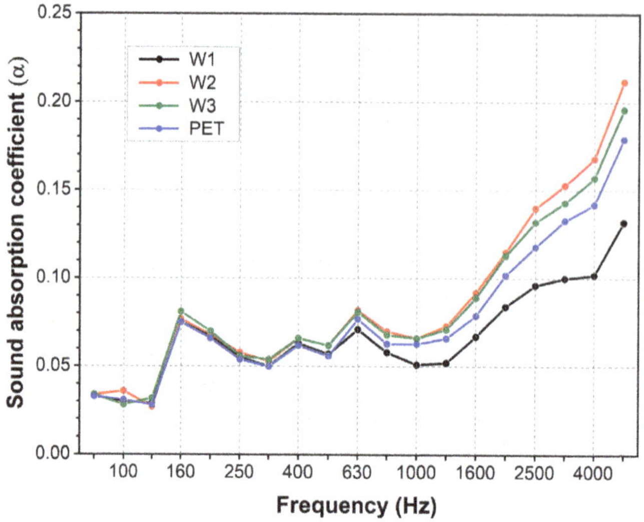

Figure 4. Sound absorption of wool and polyester nonwovens designed for carpet underlayment.

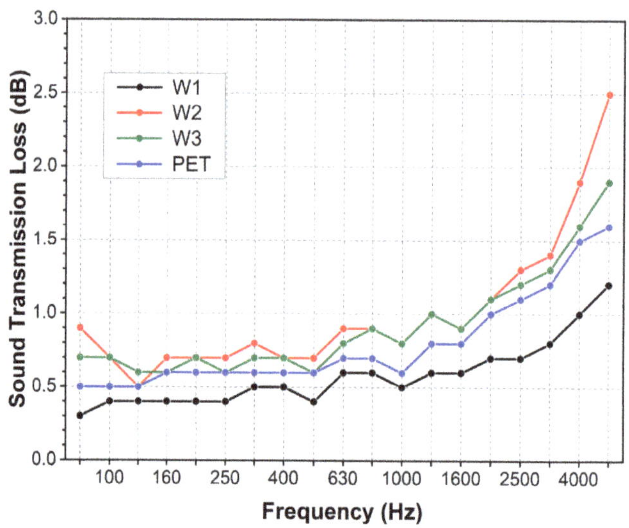

Figure 5. Sound transmission loss of wool and polyester nonwovens designed for carpet underlayment.

3.2. Sound Absorption of Carpets with Cut and Loop Piles

For carpets with cut piles, the absorption coefficients accomplished values of 0.04 to 0.9 (Figure 6). In the range of low frequencies until 250 Hz, for all samples with and without the underlayment, the SAC achieves a low value below 0.1. At medium frequencies from 250 to 1600 Hz, the SAC takes on increasingly larger values, reaching a value of 0.4 at the end of this range. In the first range of medium frequencies from 250 to 630 Hz, the increments in the SAC are minimal for the carpets containing only the primary (PB) and secondary backing (SB). For samples coupled with underlayment, independently of the underlayment type, the increments in the SAC in this range are about 0.1 higher. A further increase in the SAC is observed for higher frequencies above 1600 Hz for all samples. For samples coupled with the underlayment, the SAC achieves a maximum of 0.85. The SAC value (0.9) is slightly higher for samples containing only primary and secondary backings.

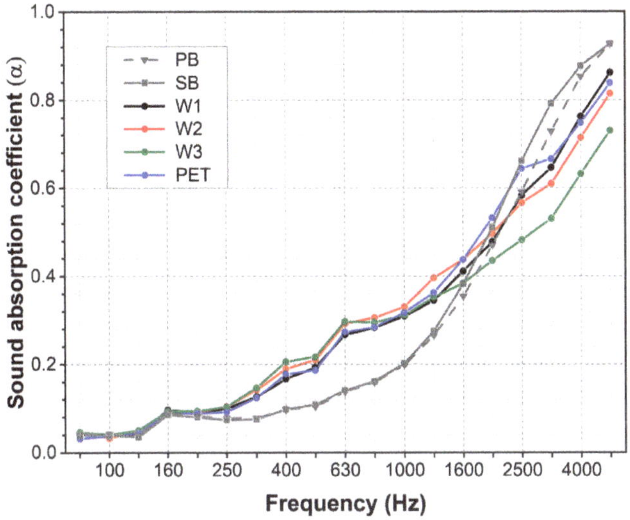

Figure 6. Sound absorption of carpets with cut piles. (PB: primary backing; SB: secondary backing).

For loop pile carpets, the values of the absorption coefficients are very similar to those measured for carpets with cut piles (Figure 7). At low frequencies, the SAC achieves a value below 0.1, and then, with increasing frequency, it reaches increasingly higher values up to 0.9 for the highest frequency of 5000 Hz. Similarly, as for cut pile samples, the SAC of samples coupled with the underlayment is higher at medium frequencies. The addition of the underlayment caused the SAC values to be approximately 0.1 higher in this range.

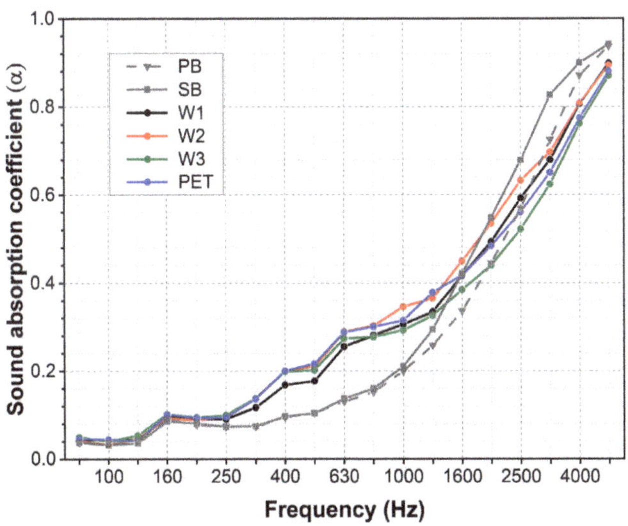

Figure 7. Sound absorption of carpets with loop piles.

For carpets with cut piles containing only the primary backing, the sound absorption average SAA equals 0.22 (Figure 8). For samples containing the secondary backing layer, the SAA is minimally higher. The SAA is approximately 25 % higher for carpets coupled with the underlayment. The SAA varies between 0.28 and 0.3 for carpets with the underlayment made from polyester and wool, independently of the nonwoven material and weight.

For carpets with loop piles formed with and without underlayment, the SAA values are practically the same as those with cut piles.

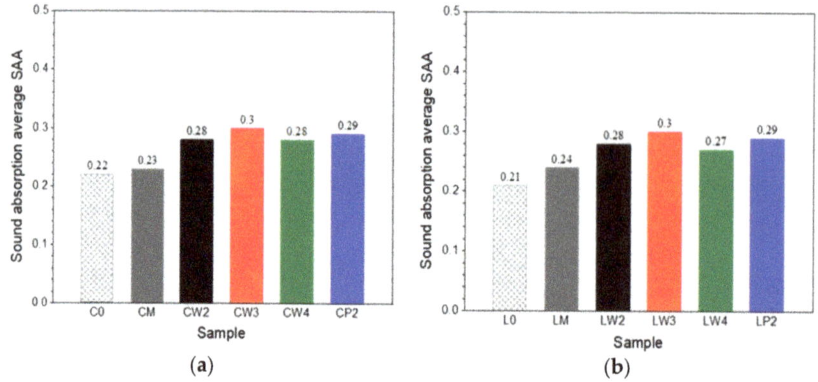

Figure 8. The sound absorption average SAA of tufted carpets with (**a**) cut pile and (**b**) loop pile.

3.3. *Sound Transmission Loss of Carpets with Cut and Loop Piles*

The sound transmission loss (STL) for carpets with cut piles varies between 5 and 25 dB (Figure 9). The STL is low for the sample with only the primary backing, reaching merely 5 dB for almost all frequencies. The STL is significantly higher for the sample with the base built from primary and secondary backings. In this case, for the low and medium frequencies, the STL fluctuates around 8 dB and then, at higher and higher frequencies, it becomes larger and larger, reaching 20 dB at the highest frequency of 5000 Hz. The STL of carpets coupled with the underlayment for low and medium frequencies varies between 12 and 18 dB. It becomes higher at higher frequencies and achieves a maximal value of 25 dB at the highest frequency.

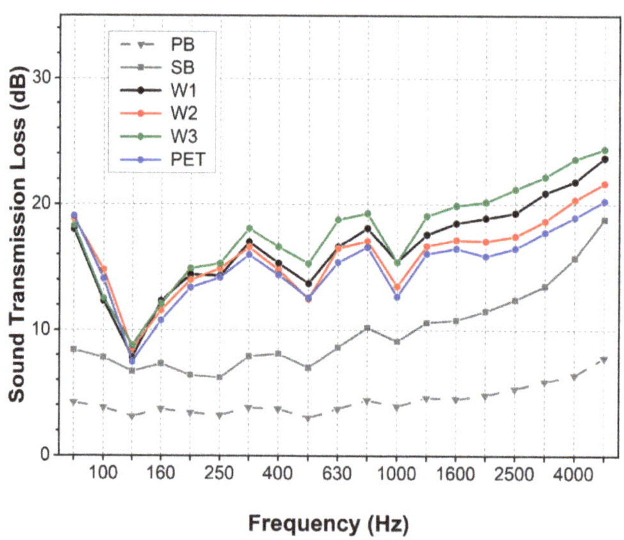

Figure 9. Sound transmission loss of carpets with cut piles.

For all frequencies, the STL of carpets coupled with the underlayment is approximately 10 dB higher than the samples containing only the carpet base. The STL values measured for carpets with different underlayments at low and medium frequencies are identical, independent of the underlayment material and unit weight. The STL is ca. 5 dB higher for wool underlayment samples at higher frequencies.

Similarly to cut-pile carpets, the STL measured for carpets with loop piles varies between 5 and 25 dB (Figure 10). Analogously, the STL is very low for the sample containing only the primary backing. The STL reaches values between 10 and 20 dB for the sample with the base constructed from primary and secondary backings. This value is higher than the STL for carpets with cut piles.

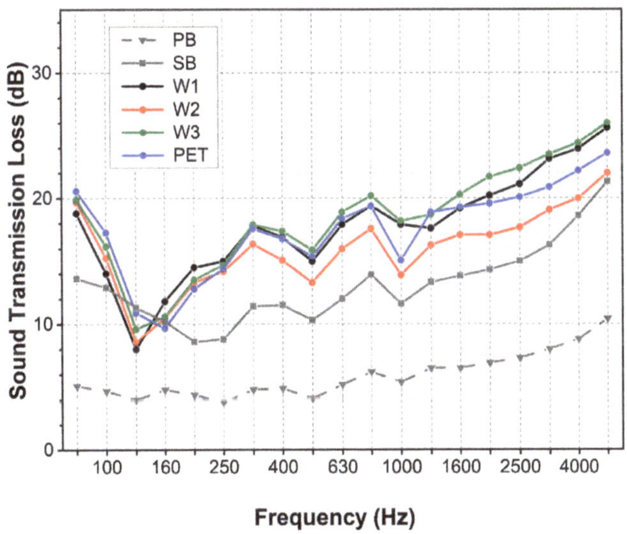

Figure 10. Sound transition loss of carpets with loop piles.

For carpets coupled with the additional underlayment, the STL varies between 12 and 18 dB at the low and medium frequencies. For these samples, the STL has the highest value of 25 dB at the highest frequency. These values are approximately 15 dB higher than values determined for the carpet with primary backing, and 10 dB higher compared to the carpet with two backings.

Regardless of the type of material, the STL of the carpets coupled with different underlayments has the same value for low and medium frequencies. Similarly to the cut piles, at higher frequencies, for samples containing the wool underlayment, the STL is ca. 5 dB higher than for the sample with two backings, and 10 dB higher than for the sample with primary backing.

4. Discussion

For carpet underlayment, the nonwovens made from wool and polyester were applied. The nonwovens were manufactured using a needle-punching technique commonly used to produce nonwoven materials. During manufacturing, the punched needles pass through a voluminous fibrous web, forming a compact structure of mechanically entangled and interlocked fibres. For both types of nonwovens, their structure consists of many built-in tortuous capillary channels and interconnected pores, which provide the nonwoven with high accessibility to the sound wave. The incidental sound waves entering and passing the nonwoven cause the air molecules accumulated within pores to vibrate. As a result of friction, the oscillating air molecules lose their energy, transforming the sound wave energy into thermal and viscous heat.

Like typical porous absorbers, the nonwovens reveal their lowest absorption capacity at low and medium sound wave frequencies [31]. Part of this range of frequencies, from 60 to 250 Hz, falls on the normal speaking vocal range [32]. For high frequencies, the energy loss of sound waves in the nonwovens occurs more intensively, so the absorption coefficients achieve higher values.

The determined coefficients for the investigated nonwovens are relatively low, and lower than those obtained for needle-punched nonwovens manufactured from wool or polyester fibres examined in other studies [33,34]. It is known that the sound absorption capacity of the nonwovens depends on several parameters, including manufacturing method, fibre characteristics, thickness, mass per unit area, porosity, tortuosity, and other structural parameters [35–38]. The low absorption of the investigated nonwovens resulted from their less-dense structure and relatively low thickness.

The wool nonwovens' absorption coefficients across the whole frequency range are similar to those determined for the nonwoven made from polyester fibres. Despite the essential differentiation in surface characteristics between wool and polyester fibres, the sound absorption capacity for both nonwovens is indistinguishable, regardless of the fibre type. Their efficiency in sound absorption is also almost identical despite the minor discrepancies in the nonwovens' thickness and mass per unit area.

Across the whole frequency range, the sound transmission loss of nonwovens achieved very low values. The sporadic papers on the transmission loss of nonwovens report that sound transmission loss is most influenced by the weight of the initial carded web and increases with its increasing weight [39]. The investigated nonwovens actually possess low weight, and, due to low needle-punching density and low thickness, their ability to insulate sound is low.

Adding the underlayment to the pile carpets' base layer forms a three-layered structure, significantly changing the carpets' sound absorption capacity and considerably influencing its sound transmission. The underlayment possesses a structure and geometry different from the piles' and base layers' and reacts differently to incident sound waves. A random web of fibres in the nonwovens forming the underlayment creates more tortuous channels, increasing the sound absorption and reducing the sound-wave energy transmitted through the material more effectively. Consequently, carpets' acoustic performance strongly depends on the characteristics of separate layers.

Coupling the underlayment obviously increases the carpet thickness, increasing the path length along which the sound wave travels through the carpet. Due to the extension of the path, more sound energy is converted into thermal and viscous heat. The enhancement of sound absorption by increasing material thickness by adding nonwoven layers is a common practice and was repeatedly observed for multilayered structures [40–42].

Increasing the carpet's thickness by coupling an underlayment does not impact the carpet's absorption capacity at low sound-wave frequencies. In the low frequencies range, the sound absorption of carpets coupled with the underlayment is the same as that of carpets without underlayment. The positive impact of the underlayment on sound absorption is observed at medium frequencies. The absorption coefficients determined for carpets coupled with the underlayment are significantly greater in this range. This is observed for underlayments made from wool and polyester. For the higher frequencies, the impact of the underlayment on the absorption capacity becomes again imperceptible. For all samples with and without underlayment, the absorption coefficients were high and only slightly differed from each other.

While the underlayment's influence on carpet sound absorption is observed only at medium frequencies, its impact on transmission loss was observed across the whole frequency range. A significant increase in transmission loss across the whole frequency range was observed for carpets with a base constructed with two backing layers. Adding the thin second backing for carpets with cut and loop piles increases transmission loss by 5 and 8 dB, respectively. The coupling of the additional underlayment leads to a further increment of the transmission loss, for carpets with both cut and loop piles.

The observed increase in the STL for a carpet made of three layers is close to 10 dB. Such an increase caused by the addition of the underlayment is much greater than the STL of the nonwovens used as carpet underlayment alone. The cumulative STL of the carpet is greater than the sum of the STLs for the individual layers.

The transmission loss of carpets, coupled with the underlayment, reached 20 dB. Compared to other studies, this level is two times smaller than the transmission loss of the composite acoustic panels, but it is two times greater than that of the assembly of loose wool or cotton fibres [43].

A similar increase in transmission loss is observed for carpets coupled with each underlayment, regardless of the fibre type and nonwoven parameters.

5. Conclusions

Applying a nonwoven as carpet underlayment provides better sound control in the room. The underlayment improves the carpet's sound absorption only at medium sound wave frequencies. At low and high frequencies, the impact of the underlayment on the carpet's sound absorption is invisible. The coupling of the underlayment significantly improves the sound insulation. The transmission loss is almost two times greater for carpets coupled with the underlayment across the whole frequency range. Carpets with an underlayment made from a wool nonwoven may efficiently attenuate noise and reduce its transmission across the floor.

The acoustic performance of the nonwoven made from waste wool obtained from the tannery industry was the same as that of the polyester nonwoven. Adding wool underlayment improves the acoustic characteristics of the carpets to the same extent as the underlayment made from polyester.

Considering acoustic performance, wool nonwovens can be an effective, eco-friendly, sound-absorbing carpet underlayment, a reasonable alternative to commonly used products from synthetic fibres. Under the zero-waste strategy, using wool for manufacturing underlayments may reduce the amount of wool waste and improve its utilisation.

Using wool underlayment enables the manufacturing of carpets in which the two layers, the pile layer and underlayment, are made from sustainable natural fibres. In this way, synthetic fibres are limited to a thin layer of the carpet base, which significantly contributes to the decrease in the pollution of the environment with plastic residues.

The use of nonwoven fabric made from tannery wool waste may involve certain inconveniences resulting from the unique characteristics of the wool. Potential disadvantages, such as low moth resistance, are a challenge that can be overcome with further research.

Author Contributions: Conceptualization, J.B. and K.K.-M.; methodology, M.B.; validation, J.B., K.K.-M. and M.R.; formal analysis, J.B. and M.B.; investigation, M.B. and K.K.-M.; resources, J.B.; data curation, K.K.-M.; writing—original draft preparation, J.B.; writing—review and editing, J.B.; visualisation, M.B. and M.R.; supervision, J.B.; project administration, J.B.; funding acquisition, M.R. All authors have read and agreed to the published version of the manuscript.

Funding: This research was funded by Norway Grants 2014-2021 through the Polish National Centre for Research and Development in the frame of the Applied Research Programme (NOR/POLNOR/WOOLUME/0007/2019-00).

Institutional Review Board Statement: Not applicable.

Informed Consent Statement: Not applicable.

Data Availability Statement: The raw data supporting the conclusions of this article will be made available by the authors on request.

Conflicts of Interest: The authors declare no conflicts of interest. The funders had no role in the study's design; in the collection, analyses, or interpretation of data; in the writing of the manuscript; or in the decision to publish the results.

References

1. Garg, N. *Environmental Noise Control: The Indian Perspective in an International Context*; Springer Nature: Berlin/Heidelberg, Germany, 2022.
2. Paul, P.; Mishra, R.; Behera, B.K. Acoustic Behaviour of Textile Structures. *Text. Prog.* **2021**, *53*, 1–64. [CrossRef]
3. Cazzolato, B.; West, C.; Schembri, T.; Watkins, P.; Robertson, W. Enhancing Acoustic Comfort with Window Coverings: Reducing Sound Transmission and Reverberation Times with a Single Product. *J. Acoust. Soc. Am.* **2023**, *154*, A127. [CrossRef]
4. Del Rey, R.; Alba, J.; Blanes, M.; Marco, B. The Acoustic Absorption of Textile Curtains on the Function of the Fullness. *Mater. Constr.* **2013**, *63*, 569–580. [CrossRef]
5. Kumar, S.; Aow, J.W.; Dexuan, W.; Lee, H.P. Investigation of Lightweight Acoustic Curtains for Mid-to-High Frequency Noise Insulations. *arXiv* **2021**, arXiv:2108.10683 2021.
6. Küçük, M.; Korkmaz, Y. Acoustic and Thermal Properties of Polypropylene Carpets: Effect of Pile Length and Loop Density. *Fibers Polym.* **2019**, *20*, 1519–1525. [CrossRef]
7. Shoshani, Y.; Wilding, M. Effect of Pile Parameters on the Noise Absorption Capacity of Tufted Carpets. *Text. Res. J.* **1991**, *61*, 736–742. [CrossRef]
8. Küçük, M.; Korkmaz, Y. Sound Absorption Properties of Acrylic Carpets. *J. Text. Inst.* **2017**, *108*, 1398–1405. [CrossRef]
9. Ardekani, T.; Yazdi, A.A.; Vadood, M.; Garai, M. Measuring and Modeling the Effect of Density and Pile Height on Sound Absorption of Double Base Persian Rug. *J. Ind. Text.* **2022**, *51*, 5728S–5755S. [CrossRef]
10. Maleki, H.; Rahbar, R.S.; Zolfaghari, S.A. Effects of Weaving Parameters on Acoustic and Thermal Insulation Properties of Handmade Carpets. *J. Text. Inst.* **2024**, *115*, 248–258. [CrossRef]
11. Samuel, B.T.; Barburski, M.; Witczak, E.; Jasińska, I. The Influence of Physical Properties and Increasing Woven Fabric Layers on the Noise Absorption Capacity. *Materials* **2021**, *14*, 6220. [CrossRef] [PubMed]
12. Ardekani, T.; Alamdar-Yazdi, A.; Vadood, M.; Garai, M. Measuring and Modelling the Effect of Base Zone on Sound Absorption of Persian Rug. *J. Text. Inst.* **2021**, *113*, 2778–2786. [CrossRef]
13. Elkhateeb, A.; Adas, A.; Attia, M.; Balila, Y. Absorption Characteristics of Masjid Carpets. *Appl. Acoust.* **2016**, *105*, 143–155. [CrossRef]
14. McCall, R.; McNeil, S. Comparison of the Energy, Time and Water Usage Required for Maintaining Carpets and Hard Floors. *Indoor Built Environ.* **2007**, *16*, 482–486. [CrossRef]
15. Ryder, M. Ancient Technique's Diverse Designs. *Wool Record* **2001**, *160*, 3681.
16. Serrano, A.; Meijer, S.; van Rijn, R.R.; Coban, S.B.; Reissland, B.; Hermens, E.; Batenburg, K.J.; van Bommel, M. A Non-Invasive Imaging Approach for Improved Assessments on the Construction and the Condition of Historical Knotted-Pile Carpets. *J. Cult. Herit.* **2021**, *47*, 79–88. [CrossRef]
17. Wood, E.J. Developments in Wool Carpet Manufacture. In *Advances in Carpet Manufacture*; Woodhead Publishing: Sawston, UK, 2009.
18. Ingham, P.; McNeil, S.; Meade, W.; Sunderland, M. Wool and Carpets—6000 Years of Innovation, Quality and Sustainability. *Key Eng. Mater.* **2016**, *671*, 490–496. [CrossRef]
19. Ansari-Renani, H.R. Fiber Quality of Iranian Carpet-Wool Sheep Breeds. *Media Peternak.* **2012**, *35*, 179–184. [CrossRef]
20. Dashab, G.; Edriss, M.; Aghaji, A.G.; Movasagh, H.; Nilforoosh, M. Wool Fiber Quality of Naeini Sheep. *Pak. J. Biol. Sci.* **2006**, *9*, 270–276. [CrossRef]
21. Haugrønning, V.; Broda, J.; Espelien, I.S.; Klepp, I.G.; Kobiela-Mendrek, K.; Rom, M.; Sigaard, A.S.; Tobiasson, T.S. Upping the WOOLUME: Waste Prevention Based on Optimal Use of Materials. In *Local, Slow and Sustainable Fashion: Wool as a Fabric for Change*; Springer: Cham, Switzerland, 2022.
22. Jakubowska, A.K.; Villavicencio, A.M.; Zimniewska, M.; Przybylska, P.; Kwiatkowska, E. Evaluation of Wool Quality Parameters of Polish Sheep Breeds. *J. Nat. Fibers* **2021**, *19*, 5880–5887. [CrossRef]
23. Berg, L.L.; Klepp, I.G.; Sigaard, A.S.; Broda, J.; Rom, M.; Kobiela-Mendrek, K. Reducing Plastic in Consumer Goods: Opportunities for Coarser Wool. *Fibers* **2023**, *11*, 15. [CrossRef]
24. Broda, J.; Kobiela-Mendrek, K.; Bączek, M.; Rom, M.; Espelien, I. Sound Absorption of Tufted Carpets Produced from Coarse Wool of Mountain Sheep. *J. Nat. Fibers* **2023**, *20*, 2246103. [CrossRef]

25. Kobiela-Mendrek, K.; Bączek, M.; Broda, J.; Rom, M.; Espelien, I.; Klepp, I. Acoustic Performance of Sound Absorbing Materials Produced from Wool of Local Mountain Sheep. *Materials* **2022**, *15*, 3139. [CrossRef]
26. Olfa, A.; Taoufik, H.; Riadh, Z.; Slah, M. The Valorization Potential of Tannery Wool Waste in the Textile Industry. *J. Nat. Fibers* **2023**, *20*, 2146251. [CrossRef]
27. Nassef, M.; Elbasyoni, A.M.; A Badr, A.; Alnahrawy, A.; Hassanin, A.H. Manufacturing and Utilization of Novel Sustainable Composites Using Pulled Wool Fibers Waste from Leather Tanneries: Mechanical, Physical, and Dynamic Characterization. *J. Ind. Text.* **2022**, *51*, 5708S–5727S. [CrossRef]
28. Abdellaoui, O.; Harizi, T.; Msahli, S. Effect of the Chemical Unhairing Process on Pulled Wool Characteristics. *Fibres Text. East. Eur.* **2022**, *30*, 70–78. [CrossRef]
29. Broda, J.; Gawlowski, A.; Rom, M.; Kobiela-Mendrek, K. Utilisation of Waste Wool from Mountain Sheep as Fertiliser in Winter Wheat Cultivation. *J. Nat. Fibers* **2023**, *20*, 2200047. [CrossRef]
30. Kicińska-Jakubowska, A.; Broda, J.; Zimniewska, M.; Przybylska, P. Characteristics of Wool of Selected Polish Local Sheep Breeds with Mixed Type Fleece. *J. Nat. Fibers* **2024**, *21*, 2290855. [CrossRef]
31. Jafari, M.J.; Monazam, M.R.; Kazempour, M. Providing an Optimal Porous Absorbent Pattern to Reduce Mid to Low-Frequency Sounds. *J. Environ. Health Sci. Eng.* **2018**, *16*, 289–297. [CrossRef] [PubMed]
32. Berg, M.; Fuchs, M.; Wirkner, K.; Loeffler, M.; Engel, C.; Berger, T. The Speaking Voice in the General Population: Normative Data and Associations to Sociodemographic and Lifestyle Factors. *J. Voice* **2017**, *31*, 257.e13–257.e24. [CrossRef] [PubMed]
33. Palak, H.; Kayaoğlu, B.K. Analysis of the Effect of Fiber Cross Section and Different Bonding Methods on Sound Absorption Performance of PET Fiber Based Nonwovens Using Taguchi Method. *J. Text. Inst.* **2020**, *111*, 575–585. [CrossRef]
34. Broda, J.; Bączek, M. Acoustic Properties of Multi-Layer Wool Nonwoven Structures. *J. Nat. Fibers* **2020**, *17*, 1567–1581. [CrossRef]
35. Shahani, F.; Soltani, P.; Zarrebini, M. The Analysis of Acoustic Characteristics and Sound Absorption Coefficient of Needle Punched Nonwoven Fabrics. *J. Eng. Fibers Fabr.* **2014**, *9*, 155892501400900210. [CrossRef]
36. Puranik, P.R.; Parmar, R.R.; Rana, P.P. Nonwoven Acoustic Textiles—A Review. *Int. J. Adv. Res. Eng. Technol.* **2014**, *5*, 81–88.
37. Paul, P.; Ahirwar, M.; Behera, B. Acoustic Behaviour of Needle Punched Nonwoven Structures Produced from Various Natural and Synthetic Fibers. *Appl. Acoust.* **2022**, *199*, 109043. [CrossRef]
38. Gokarneshan, N. Acoustic Behaviour of Nonwovens. *J. Text. Sci. Fash. Technol.* **2020**, *5*, 1–3. [CrossRef]
39. Ghorbani, K.; Hasani, H.; Zarrebini, M.; Saghafi, R. An Investigation into Sound Transmission Loss by Polypropylene Needle-Punched Nonwovens. *Alex. Eng. J.* **2016**, *55*, 907–914. [CrossRef]
40. Liu, J.; Liu, X.; Bao, W.; Wang, S.; Chen, L.; Chen, Z.; Zuo, B.; Gao, W. The Acoustic Behaviors of Dual Layered Nonwoven Absorbers: From Model Building to Experiment Verification. *J. Comput. Acoust.* **2014**, *22*, 1450001. [CrossRef]
41. Kucukali-Ozturk, M.; Nergis, B.; Candan, C. Design of Layered Structure with Nanofibrous Resonant Membrane for Acoustic Applications. *J. Ind. Text.* **2018**, *47*, 1739–1756. [CrossRef]
42. Yilmaz, N.D.; Banks-Lee, P.; Powell, N.B.; Michielsen, S. Effects of Porosity, Fiber Size, and Layering Sequence on Sound Absorption Performance of Needle-Punched Nonwovens. *J. Appl. Polym. Sci.* **2011**, *121*, 3056–3069. [CrossRef]
43. El Messiry, M.; Ayman, Y. Investigation of Sound Transmission Loss of Natural Fiber/Rubber Crumbs Composite Panels. *J. Ind. Text.* **2022**, *51*, 5347S–5369S. [CrossRef]

materials

Communication

A Compact Low-Frequency Acoustic Perfect Absorber Constructed with a Folded Slit

Han Wang *, Pengwei Ma and Xueling Fan *

Xi'an Key Laboratory of Extreme Environment and Protection Technology, School of Aerospace Engineering, Xi'an Jiaotong University, Xi'an 710049, China
* Correspondence: whishappy@foxmail.com (H.W.); fanxueling@mail.xjtu.edu.cn (X.F.)

Abstract: Tunable perfect acoustic absorption at subwavelength thickness has been a prominent topic in scientific research and engineering applications. Although metamaterials such as labyrinthine metasurfaces and coiling-up-space metamaterials can achieve subwavelength low-frequency acoustic absorption, efficiently realizing tunable absorption under uniform and limited size conditions remains challenging. In this paper, we introduce a folded slit to enhance the micro-slit acoustic absorber, effectively improving its low-frequency acoustic absorption performance and successfully achieving a perfect acoustic absorption coefficient of 0.99 at a thickness of only 3.1 cm. By adjusting just two parameters of the folded area, we can efficiently achieve a tunable resonant frequency ranging from 525 to 673 Hz and a tunable acoustic absorption bandwidth of 56.5% to 60.2%, simultaneously maintaining uniform external dimensions. Additionally, the folded-slit absorber demonstrates a broader acoustic absorption bandwidth at lower frequencies, enhancing broadband absorption capabilities in the low-frequency domain. These results hold significant potential for the design of highly efficient, thin and tunable acoustic absorbers.

Keywords: perfect acoustic absorption; slit absorber; folded slit; tunable absorption; acoustic metamaterial

1. Introduction

The presence of noise hinders individuals' rest, study and work, leading to detrimental impacts on both human health and the surrounding environment [1]. With the development of industrial production, transportation and urban buildings, environmental noise is becoming increasingly serious. It has become a major public hazard polluting the human social environment [2,3]. In the field of aviation, aircraft engines produce intense noise during the service process, which greatly reduces the comfort of people in the aircraft [4]. Prolonged exposure to a noisy environment can have adverse effects on the health of the crew and passengers. Therefore, measures should be taken to mitigate noise effects and enhance aircraft comfort. This has important scientific and engineering application value and is also an important aspect of the future technological development of the aviation industry.

Conventional acoustic absorbers such as micro-slit absorbers [5,6], porous materials [7,8] and micro-perforated plate absorbers [9] exhibit a narrow acoustic absorption band, and their structural thickness is typically equivalent to a quarter of the working frequency wavelength [10]. Consequently, traditional absorbing materials/structures are thick when used to absorb noise, especially for low-frequency noise, and so cannot meet the technical requirements of compact-space noise reduction, such as for aircraft engines, automobiles and small rooms, at the present stage. For compact interior spaces, there are strict requirements for the volume and thickness of acoustic absorbers, which greatly limits the practical application potential of conventional absorbers. Therefore, it is very important to find a new structure with excellent acoustic absorption performance in the low-frequency range and the potential to reduce the thickness of the structural elements.

In recent years, acoustic metamaterials and metasurfaces have received a lot of attention and undergone rapid development because they present many excellent and special acoustic properties, enabling new methods and ideas for the innovation of aircraft engine noise reduction technology [11]. Acoustic absorption structures such as labyrinthine metasurfaces [12,13], folded-cavity metamaterials [14,15] and coiling-up-space metamaterials [16,17] can achieve low-frequency absorption under subwavelength thickness. However, because they have a large number of curved channels and a complex structure, it is difficult to achieve tunable acoustic absorption without changing their overall size. Therefore, it is difficult to maintain a consistent shape and achieve tunable and efficient acoustic absorption.

In the present study, we propose a novel design for an absorber with a folded slit. The folded slit was introduced by modifying the original micro-slit absorber design. The perfect acoustic absorption of the folded-slit absorber was investigated through simulations and experiments. Moreover, the folded-slit absorber efficiently achieved a tunable resonant frequency and a tunable acoustic absorption bandwidth by only adjusting the parameters of the folded area while maintaining uniform external dimensions. This approach may pave the way for highly efficient, thin and tunable acoustic absorbers.

2. Structure and Methods

2.1. Structure of Folded-Slit Absorber

The novel folded-slit absorber consists of a folded-slit panel and a cavity, as illustrated in Figure 1. The parameter t_p represents the thickness of the upper panel, while t_c denotes the depth of the cavity. The parameter l_c indicates the width of the cavity, and l_s refers to the width of the slit. In the folded area, the thickness, t_i, is n times the thickness of the upper panel, t_p (where $n = t_i/t_p$, $0 < n < 1$). Additionally, the width, l_i, of this area is m times the width of the slit, l_s (where $m = l_i/l_s$, $1 < m < l_c/l_s$). By adjusting the parameters m and n, the dimensions of the folding area can be modified, thereby enhancing the ability to manipulate the acoustic impedance of the absorber.

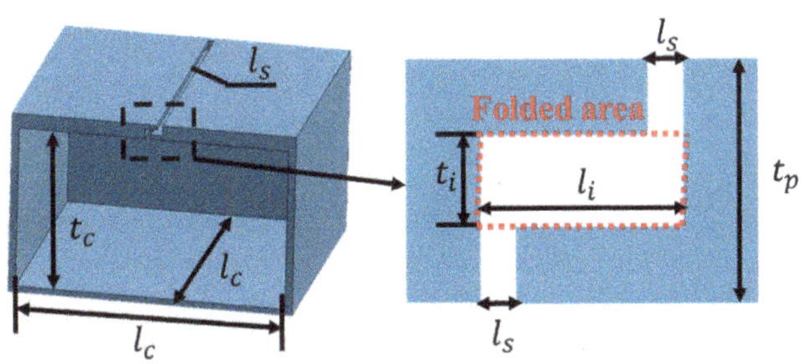

Figure 1. Structural diagram of the novel folded-slit absorber.

2.2. Method of Finite Element Simulation

The acoustic–thermoacoustic interaction models used to calculate the acoustic absorption coefficient were constructed using COMSOL 5.6 finite element software. As illustrated in Figure 2a, a plane wave is incident from above the absorber along a virtual impedance tube. The sound pressure of the plane wave, P_i, is given by

$$P_i = Ae^{iky} \tag{1}$$

where A denotes the amplitude of acoustic pressure, and $k = \omega/c_0$ is the wavenumber. The pressure, P_{y_1}, P_{y_2}, at the interface y_1, y_2 can be expressed as [18]

$$P_{y_1} = A_i e^{iky_1} + A_r e^{-iky_1} \tag{2}$$

$$P_{y_2} = A_i e^{iky_2} + A_r e^{-iky_2} \tag{3}$$

and is solved as

$$A_i = \frac{b_2 P_{y_1} - a_2 P_{y_2}}{a_1 b_2 - a_2 b_1}, A_r = \frac{a_1 P_{y_2} - b_1 P_{y_1}}{a_1 b_2 - a_2 b_1} \quad (4)$$

where $a_1 = e^{iky_1}$, $a_2 = e^{-iky_1}$, $b_1 = e^{iky_2}$ and $b_2 = e^{-iky_2}$. Thus, the absorption coefficient can be calculated by

$$\alpha = 1 - \left|\frac{A_r}{A_i}\right|^2 \quad (5)$$

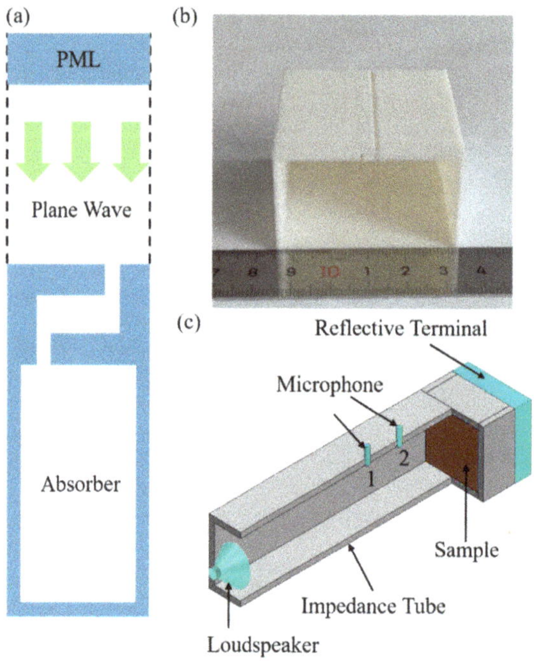

Figure 2. (**a**) The finite element model of the folded-slit absorber; (**b**) a photo of the folded-slit absorber; (**c**) a schematic illustration for the experimental realization of the designed acoustic absorbers.

A perfect matching layer (PML) was added to the top of the tube to avoid spurious refection. The wall of the structure was considered to be the sound hard boundary. The effects of viscous and heat transfer are included in the linearized Navier–Stokes equation. Boundary-layer meshes were applied in the finite element models due to the significant velocity and temperature gradients within the boundary layer. In the simulation, the following parameters were adopted: air density, ρ_0 = 1.21 kg/m^3; sound speed, $c_0 = 343$ m/s; dynamic viscosity, $\eta = 1.56 \times 10^{-5}$Pa·s; temperature, T = 293.15 K; heat capacity at constant pressure, C_P = 1.013 $\frac{\text{kJ}}{\text{kg}\cdot\text{K}}$; and thermal conductivity, $\mu = 2.593 \times 10^{-2}$ W/mk.

2.3. Experimental Method

The folded-slit absorbers in this study were fabricated by a 3D printer (Creator-pro, FLASHFORGE, Hangzhou, China) with the fused deposition modeling (FDM) technique. Wire polylactic acid (PLA, FLASHFORGE, Hangzhou, China) was used as a raw material for manufacturing specimens of the folded-slit absorbers. The corresponding 3D printing parameters were as follows: a slice thickness of 0.12 mm, a nozzle preheating temperature of 200 °C, a printing speed of 50 mm/s, a printing accuracy of ±0.4 mm, etc. After the 3D printing process, the extra support materials were removed, and the specimens were cleaned manually. The printed sample is shown in Figure 2b.

The measurements of the absorption coefficient were performed using a square impedance tube system (F50, BSWA, Beijing, China), which complies with ASTM E1050-12 standards [19]. As shown in Figure 2c, a loudspeaker was mounted on one end of the tube, with the absorber mounted on the other end. The side length of the tube cross-section was 50 mm. The sound pressure was measured separately by two microphones with a distance interval of s:

$$p_1 = A_{1I}e^{jk_0x_1} + A_{1R}e^{-jk_0x_1} \tag{6}$$

$$p_2 = A_{2I}e^{jk_0x_2} + A_{2R}e^{-jk_0x_2} \tag{7}$$

where A_I and A_R are the sound pressure amplitudes of the incident wave and the reflected wave, respectively. Then, the transfer functions can be described as [20]

$$H_I = \frac{A_{2I}}{A_{1I}} \tag{8}$$

$$H_R = \frac{A_{2R}}{A_{1R}} \tag{9}$$

$$H_{12} = \frac{p_2}{p_1} \tag{10}$$

The acoustic absorption coefficient measured in this experiment is

$$\alpha = 1 - \left|\frac{H_{12} - H_I}{H_R - H_{12}}e^{j2ks}\right|^2 \tag{11}$$

3. Results and Discussion

3.1. Low-Frequency Perfect Acoustic Absorption of the Folded-Slit Absorber

Figure 3 compares the acoustic absorption coefficients of the folded-slit absorber with those of a traditional micro-slit absorber. The specific structural parameters for both types are detailed in Table 1. The experimental results indicate that the micro-slit absorber achieves its peak absorption at 721 Hz, with an absorption coefficient of 0.91. Notably, the cavity thickness of this absorber is 1/17 of the wavelength corresponding to the resonant frequency. In contrast, when the thickness and width of the folded-slit absorber are kept constant, its resonant frequency, f_r, is measured to be 600 Hz, which is 16.7% lower than that of the micro-slit absorber. Furthermore, the cavity thickness of the folded-slit absorber is just 1/21 of the resonant frequency wavelength, but it attains a peak absorption coefficient of 1.0, signifying perfect acoustic absorption. These findings demonstrate that incorporating the folded slit effectively enhances low-frequency performance and facilitates the design of a perfect absorber at subwavelength scales.

Table 1. The structural parameters of the folded-slit absorber and micro-slit absorber.

Parameter	Folded-Slit Absorber	Micro-Slit Absorber	Unit
t_p	2.5	2.5	mm
t_c	27.5	27.5	mm
l_c	47.0	47.0	mm
l_s	1.0	1.0	mm
t_i	1.0	/	mm
l_i	3.0	/	mm

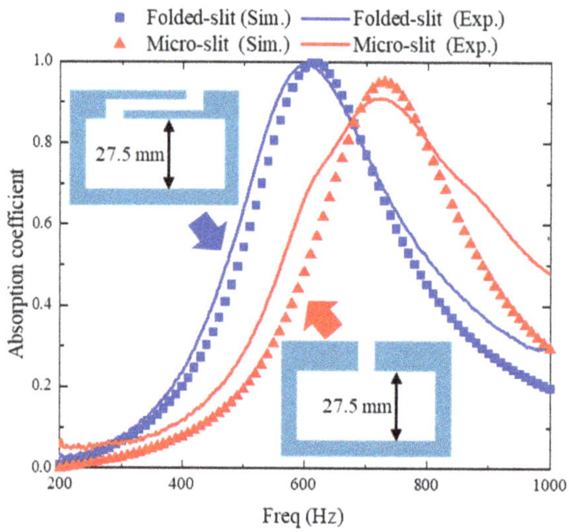

Figure 3. Performance comparison between the folded-slit absorber and micro-slit absorber.

In the simulation models, the structural walls were treated as hard sound boundaries. However, in reality, the actual material behaves as a lossy medium. This means that the sound energy loss was only approximated in the simulation models, potentially leading to discrepancies between the simulation and the measurement results. Nonetheless, a consistent conclusion can be drawn from comparing the simulation results: the folded-slit absorber achieves perfect acoustic absorption and lower-frequency absorption under uniform dimensions, exhibiting superior subwavelength performance.

Figure 4a displays the air particle velocities of the folded-slit and micro-slit absorbers at their resonant frequencies. In both structures, the air particle velocities were significantly higher within the slits than elsewhere. At the resonant frequencies, the vigorous movement of air particles in the area formed by the slits resulted in peak velocities, akin to a resonant cavity dominated by sound mass. Conversely, the slit region of the micro-slit absorber is a straight channel, resulting in less intense particle vibration compared to the folded-slit absorber. The folded design prolongs the sound wave propagation path, increasing thermal viscosity loss due to the slits. Based on the relationship between the acoustic absorption coefficient and acoustic impedance,

$$\alpha = 1 - \left| \frac{Z_s - \rho_0 c_0}{Z_s + \rho_0 c_0} \right|^2, \tag{12}$$

when the acoustic reactance, $\mathrm{Im}(Z_s)$, is 0 and the acoustic resistance, $\mathrm{Re}(Z_s)$, approaches 1, the acoustic absorber demonstrates higher absorption efficiency. As shown in Figure 4b, the acoustic reactance of the micro-slit sound absorber nears 0 at the resonant frequency; however, the straight slit does not provide sufficient acoustic resistance, preventing perfect acoustic absorption. The folded-slit absorber enhances structural resistance, allowing for impedance matching at the resonance frequency, thus achieving perfect absorption. This further illustrates the effectiveness of employing folded slits in designing low-frequency acoustic absorbers.

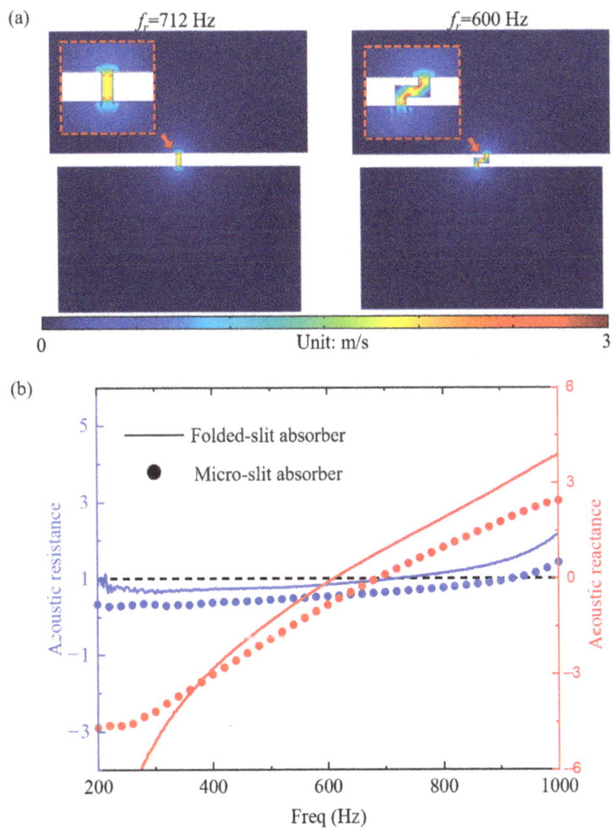

Figure 4. The absorption mechanism of the folded-slit absorber. (**a**) The particle velocities and (**b**) the acoustic resistance and reactance of the folded-slit and micro-slit absorbers at their resonant frequencies.

3.2. Tunable Perfect Absorption of the Folded-Slit Absorber

When its overall dimensions are fixed, an absorber with tunable perfect acoustic absorption offers significant benefits for noise reduction across different contexts and facilitates the design of multi-unit coupled broadband acoustic absorption systems [21]. Therefore, a tunable perfect absorber is a crucial component for practical applications. Labyrinthine metasurfaces face challenges in achieving a tunable absorption bandwidth with a consistent shape due to the necessity of increasing the number of channels [12,13]. Conversely, coiling-up-space metamaterials [16,17] and folded-cavity metamaterials [14,15] can adjust their absorption frequencies while retaining the same dimensions. However, they involve complex bending channels and numerous structural parameters that complicate performance tuning. Thus, designing a tunable perfect absorber with constant dimensions remains a significant challenge. In the following, we demonstrate how folded-slit absorbers can be effectively tuned while maintaining a constant external shape by adjusting the folded slits.

Four samples with identical dimensions were fabricated, as depicted in Figure 5a. Each sample has a thickness of 31 mm and a width of 49.6 mm. The specific parameters are detailed in Table 2. Figure 5b illustrates a comparison of the absorption coefficients for the different samples. Notably, the folded area width, l_i, of sample F2 is larger than that of sample F1. Sample F1 exhibits a resonant frequency of 600 Hz, while sample F2 has a resonant frequency of 561 Hz. These results indicate that increasing the ratio, m, of the folded area width, l_i, to the slit width, l_s, enhances the low-frequency performance of the structure.

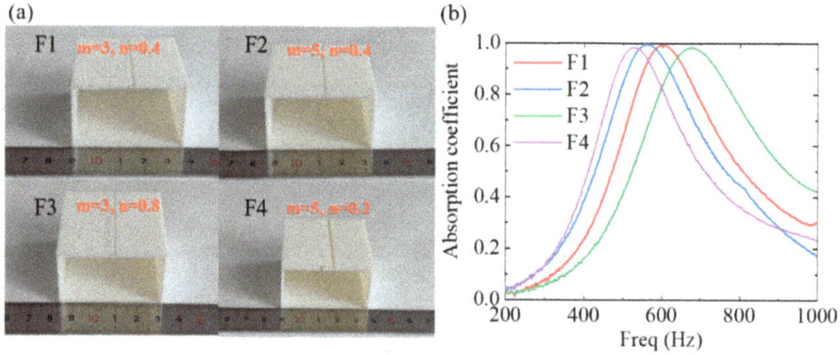

Figure 5. The tunable absorption of the folded-slit absorber. (**a**) Photos of the different samples; (**b**) the absorption coefficients of samples F1–F4.

Table 2. The structure parameters of the samples.

Parameter	F1	F2	F3	F4	Unit
t_p	2.5	2.5	2.5	5.0	mm
t_c	27.5	27.5	27.5	25.0	mm
l_c	47.0	47.0	47.0	47.0	mm
l_s	1.0	1.0	1.0	1.5	mm
t_i	1.0	1.0	2.0	1.0	mm
l_i	3.0	5.0	3.0	7.5	mm
m	3.0	5.0	3.0	5.0	/
n	0.4	0.4	0.8	0.2	/

In comparison to sample F3, sample F1 has a smaller folded area thickness, t_i. The absorption coefficient comparison shows that the resonant frequency of sample F3 is 673 Hz, which represents a shift of 73 Hz to higher frequencies compared to sample F1. A smaller ratio, n, between the thickness, t_i, of the folded area and the thickness, t_p, of the upper panel facilitates low-frequency acoustic absorption in the folded-slit absorber.

For traditional micro-slit absorbers, achieving high absorption efficiency necessitates a thin upper panel thickness and a small slit width. When the slit width exceeds 1 mm, the acoustic absorption performance declines significantly, which leads to a demand for higher manufacturing precision and increases production costs [22,23]. The folded-slit absorber sample F4 designed in this study features a substantial panel thickness (5 mm) and slit width (1.5 mm) while maintaining an efficient absorption coefficient of greater than 0.98. The resonant frequency of sample F4 is 525 Hz, with its cavity thickness being only 1/26 of the resonant wavelength, thereby simplifying production and enhancing low-frequency performance. Additionally, sample F4 shares the same parameter m as sample F2, but with a smaller parameter n value, further indicating that a reduced n value promotes effective low-frequency acoustic absorption. The absorption bandwidth [21] of the acoustic absorber can be calculated as

$$B = \frac{\Delta f}{f_r} \tag{13}$$

where Δf represents the absorption band, with an acoustic absorption coefficient of greater than 0.5, and f_r is the resonant frequency. The absorption bandwidths for samples F1 to F4 are 56.5%, 60.2%, 59.1% and 57.3%, respectively. In structures such as neck-extended absorbers, a lower resonant frequency typically correlates with a narrower absorption bandwidth under the constraints of perfect absorption and a constant external shape.

However, in contrast to sample F1, samples F2 and F4 demonstrate a broader absorption bandwidth while achieving lower frequencies for perfect acoustic absorption, effectively overcoming this limitation.

Overall, the folded-slit absorber proposed in this study can efficiently achieve tunable perfect acoustic absorption under the conditions of constant external dimensions through merely adjusting the folded area parameters. It exhibits superior acoustic adjustment performance compared to previous studies on labyrinthine metasurfaces, coiling-up-space metamaterials, and folded-cavity metamaterials. This structure design lays the foundation for broadband absorption designs.

4. Conclusions

Achieving perfect low-frequency absorption with structures of subwavelength thickness poses a challenge for conventional acoustic absorbers. This paper presents the design of an acoustic absorber featuring a folded slit, demonstrating perfect absorption both through simulations and experiments. The conclusions are as follows:

- The folded-slit absorber achieves perfect acoustic absorption when the cavity thickness is only 1/21 of the resonant frequency's wavelength. For the same dimensions, the resonant frequency is shifted down by 16.7% compared to traditional micro-slit absorbers, enhancing its performance in subwavelength low-frequency acoustic absorption.
- By adjusting the folded area parameters, m and n, the folded-slit absorber can achieve perfect acoustic absorption at multiple frequencies ranging from 525 to 673 Hz without altering its overall dimensions, demonstrating efficient tunability.
- With uniform dimensions, the acoustic absorption bandwidth broadens as the folded-slit absorber targets lower frequencies, facilitating the realization of broadband absorption in the low-frequency domain.

In summary, this paper proposes a compact folded-slit absorber that enhances the subwavelength low-frequency performance of micro-slit absorbers. With a thickness of only 3.1 cm, the folded-slit absorber efficiently achieves tunable perfect acoustic absorption. The proposed structure may contribute significantly to the design of highly efficient, ultrathin, and widely tunable acoustic absorbers, as well as to furthering broadband absorption design.

Author Contributions: Conceptualization, H.W.; data curation, H.W. and P.M.; formal analysis, H.W. and P.M.; funding acquisition, H.W. and X.F.; investigation, H.W. and P.M.; methodology, H.W.; project administration, X.F.; validation, H.W. and P.M.; visualization, H.W. and P.M.; writing—original draft, H.W.; writing—review and editing, H.W., P.M. and X.F. All authors have read and agreed to the published version of the manuscript.

Funding: This research was funded by the National Science and Technology Major Project (Grant No. J2019-IV-0003-0070), the AECC Independent Innovation Fund (Grant No. ZZCX-2023-004), the National Natural Science Foundation of China (Grant No. 12102335) and the China Postdoctoral Science Foundation (Grant No. 2021M702604).

Institutional Review Board Statement: Not applicable.

Informed Consent Statement: Not applicable.

Data Availability Statement: The raw data supporting the conclusions of this article will be made available by the authors on request.

Conflicts of Interest: The authors declare that they have no known competing financial interests or personal relationships that could have appeared to influence the work reported in this paper.

References

1. Brink, M.; Schäffer, B.; Vienneau, D.; Foraster, M.; Pieren, R.; Eze, I.C.; Cajochen, C.; Probst-Hensch, N.; Röösli, M.; Wunderli, J. A survey on exposure-response relationships for road, rail, and aircraft noise annoyance: Differences between continuous and intermittent noise. *Environ. Int.* **2019**, *125*, 277–290. [CrossRef] [PubMed]
2. Baudin, C.; Lefèvre, M.; Babisch, W.; Cadum, E.; Champelovier, P.; Dimakopoulou, K.; Houthuijs, D.; Lambert, J.; Laumon, B.; Pershagen, G.; et al. The role of aircraft noise annoyance and noise sensitivity in the association between aircraft noise levels and hypertension risk: Results of a pooled analysis from seven European countries. *Environ. Res.* **2020**, *191*, 110179. [CrossRef]
3. Gjestland, T. A Systematic Review of the Basis for WHO's New Recommendation for Limiting Aircraft Noise Annoyance. *Int. J. Environ. Res. Public Health* **2018**, *15*, 2717. [CrossRef] [PubMed]
4. Ma, P.; Wang, H.; Li, D.; Yuan, S.; Fan, X. Passive acoustic liners for aero-engine noise control: A review. *Mech. Adv. Mater. Struct.* **2024**, 1–16. [CrossRef]
5. Liu, Q.; Zhang, C. Broadband and low-frequency sound absorption by a slit-perforated multi-layered porous metamaterial. *Eng. Struct.* **2023**, *281*, 115743. [CrossRef]
6. Ma, F.; Liu, C.; Li, Y.; Wu, J.H. A thin-walled mesoscopic hybrid slit-resonator metamaterial absorber. *Thin-Walled Struct.* **2023**, *193*, 111226. [CrossRef]
7. Berardi, U.; Iannace, G. Acoustic characterization of natural fibers for sound absorption applications. *Build. Environ.* **2015**, *94*, 840–852. [CrossRef]
8. Cao, L.; Fu, Q.; Si, Y.; Ding, B.; Yu, J. Porous materials for sound absorption. *Compos. Commun.* **2018**, *10*, 25–35. [CrossRef]
9. Maa, D. Potential of microperforated panel absorber. *J. Acoust. Soc. Am.* **1998**, *104*, 2861–2866. [CrossRef]
10. Ma, F.; Wang, C.; Du, Y.; Zhu, Z.; Wu, J.H. Enhancing of broadband sound absorption through soft matter. *Mater. Horiz.* **2022**, *9*, 653–662. [CrossRef]
11. Gao, N.; Zhang, Z.; Deng, J.; Guo, X.; Cheng, B.; Hou, H. Acoustic Metamaterials for Noise Reduction: A Review. *Adv. Mater. Technol.* **2022**, *7*, 2100698. [CrossRef]
12. Liu, L.; Chang, H.; Zhang, C.; Hu, X. Single-channel labyrinthine metasurfaces as perfect sound absorbers with tunable bandwidth. *Appl. Phys. Lett.* **2017**, *111*, 083503. [CrossRef]
13. Pavan, G.; Singh, S. Near-perfect sound absorptions in low-frequencies by varying compositions of porous labyrinthine acoustic metamaterial. *Appl. Acoust.* **2022**, *198*, 108974. [CrossRef]
14. Yang, M.; Chen, S.; Fu, C.; Sheng, P. Optimal sound-absorbing structures. *Mater. Horizons* **2017**, *4*, 673–680. [CrossRef]
15. Boulvert, J.; Costa-Baptista, J.; Cavalieri, T.; Romero-García, V.; Gabard, G.; Fotsing, E.R.; Ross, A.; Perna, M.; Mardjono, J.; Groby, J. Folded metaporous material for sub-wavelength and broadband perfect sound absorption. *Appl. Phys. Lett.* **2020**, *117*, 251902. [CrossRef]
16. Liang, M.; Wu, H.; Chen, S.; Fu, Z.; Zhang, R.; Xing, Z. Simulated and Experimental Research on Highly Efficient Sound Absorption of Low-Frequency Broadband Acoustic Metamaterial Based on Coiled-Up Space. *Adv. Eng. Mater.* **2023**, *25*, 2301221. [CrossRef]
17. Li, Y.; Assouar, B.M. Acoustic metasurface-based perfect absorber with deep subwavelength thickness. *Appl. Phys. Lett.* **2016**, *108*, 063502. [CrossRef]
18. Ma, P.; Wang, H.; Liu, Y.; Jin, X.; Wang, F.; Fan, X. New Broadband Metasurface Acoustic Liner Design Method Based on a Schroeder Diffuser Sequence. *J. Aerosp. Eng.* **2022**, *35*, 04022047. [CrossRef]
19. *ASTM E1050-12*; Standard Test Method for Impedance and Absorption of Acoustical Materials Using a Tube, Two Microphones and a Digital Frequency Analysis System. ASTM International: West Conshohocken, PA, USA, 2012.
20. Zhao, H.; Wen, J.; Yu, D.; Wen, X. Low-frequency acoustic absorption of localized resonances: Experiment and theory. *J. Appl. Phys.* **2010**, *107*, 023519. [CrossRef]
21. Huang, S.; Fang, X.; Wang, X.; Assouar, B.; Cheng, Q.; Li, Y. Acoustic perfect absorbers via Helmholtz resonators with embedded apertures. *J. Acoust. Soc. Am.* **2019**, *145*, 254–262. [CrossRef]
22. Wu, C.C.; Loke, S.; Lu, Z.; Shrestha, M.; Khoo, B.C.; Lau, G. Transformation from Helmholtz to Membrane Resonance by Electro-Adhesive Zip of a Double-Layer Micro-Slit Acoustic Absorber (Adv. Mater. Technol. 10/2023). *Adv. Mater. Technol.* **2023**, *8*, 2370048. [CrossRef]
23. Brunskog, J.; Glebe, D.; Garza-Agudelo, D.; Nilsson, E. Absorption and scattering by perforated facings with periodic narrow slits. *J. Acoust. Soc. Am.* **2022**, *151*, 1847–1859. [CrossRef]

materials

MDPI

Article

The Influence of Diatomite on the Sound Absorption Ability of Composites

Michał Łach [1,*], Eulalia Gliścińska [2], Agnieszka Przybek [1,*] and Krzysztof Smoroń [3]

1 Materials Engineering, Faculty of Material Engineering and Physics, Cracow University of Technology, Jana Pawła II 37, 31-864 Kraków, Poland
2 Textile Institute, Faculty of Material Technologies and Textile Design, Lodz University of Technology, Zeromskiego 116, 90-924 Lodz, Poland; klata@p.lodz.pl
3 Specialized Mining Company GÓRTECH Sp. z o.o., 31-586 Kraków, Poland; k.smoron@diato.pl
* Correspondence: michal.lach@pk.edu.pl (M.Ł.); agnieszka.przybek@pk.edu.pl (A.P.)

Abstract: Diatomites are well-known mineral materials formed thousands of years ago from the skeletons of diatoms. They are found in many places around the world and have a wide range of applications. This article presents innovative research related to the possibility of using diatomite as a filler in composites to improve their sound absorption properties. The results of the study of the effect of diatomite processing (calcination) and its degree of fineness on the sound absorption coefficient of thermoplastic composites are presented. Three fractions of diatomite (0 ÷ 0.063 mm; 0.5 ÷ 3 mm; 2 ÷ 5 mm) and its variable mass proportion (0, 25, and 50 wt.%) were used. The composites were made with flax fibers as a reinforcement, polylactide as a matrix, and diatomite as an additional filler. This paper also presents the results of oxide chemical composition, diatomite mineral phase composition, morphology, and thermal conductivity coefficient of all diatomite fractions studied. In addition, the average particle size for diatomite powder was also determined. The most important of the studies was the determination of the acoustic properties of the aforementioned composites. As a result of the tests, it was found that the smallest fraction of diatomite particles and a variant without thermal treatment give the best effect in terms of sound absorption.

Keywords: diatomite; sorbent of petroleum substances; calcination; sound absorption; composite

1. Introduction

Diatomite is a sedimentary rock composed of the fossilized skeletal remains of single-cell aquatic "algaes" known as "diatoms". This unique form of silica has an elaborate honeycomb structure, peppered with thousands of tiny holes ranging from a few microns to submicron diameters. No other silica source, be it mined or artificially produced, presents such a structure. Some diatomite deposits are saltwater, but most are from freshwater sources. When ground, it results in an extremely low-density powder known as "diatomaceous earth" (DE), which has excellent absorption properties that are highly prized for filtration, agriculture, paints, plastics, cosmetics, and pharmaceuticals applications [1]. There are large, documented deposits of diatomites all over the world, with the largest amounts in the US, China, Spain, and Turkey [2]. The locations of diatomite deposits and where diatomite mining is carried out are shown in Figure 1 [3].

As shown in Figure 1, diatomites also occur in Poland. In Poland, diatomite deposits are located in Podkarpacie, with the only active deposit located in Jawornik Ruski. Diatomites have a particularly high capacity to absorb petroleum substances and can reach a sorption capacity for these substances of up to 160 wt.%. [4]. Typical diatomites containing more than 80 wt.% SiO_2 in their composition do not occur in Poland [1]. The composition of diatomites from mines in Poland is presented in articles [5,6]. The proven balance resources of diatomite rock are about 10 million tons. The diatomites mined in Poland, shown in Figure 2, are not homogeneous materials. There are different fractions in diatomite rock, and the way it is mined and processed causes the material to have slightly different contents of amorphous silica and other minerals. Nevertheless, it is finding increasing use in various industries [4].

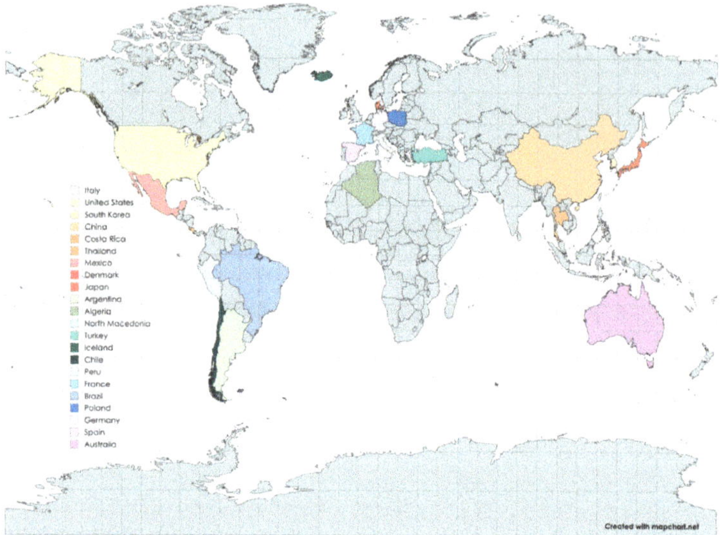

Figure 1. Map of countries where diatomite deposits are located and where diatomite mining is carried out (based on [3]).

Figure 2. Diatomite mine in Poland. Appearance of diatomite rocks–different fractions and minerals visible.

Literature reports indicate that so far, the combination of diatomite with materials such as gypsum [7], zeolite [8], magnesium oxychloride cement [9,10], bitumen [11], rubber [12–14], polyethylene glycol [15,16], sodium acetate trihydrate (SAT)-acetamide (AC)-micron/nano aluminum nitride (AlN) composite [17], a sodium silicate solution [10], chitosan [18,19], PA11 [20], polyurethane [21], polyetherimide [22], polypropylene [23], poly (lactic acid) [24], poly (methyl methacrylate) [25,26], poly (acrylamide) [27], polyaniline [28], ethylene-vinyl acetate copolymer [29], and epoxy resin [30,31] has been tested.

Diatomite can be calcined [22], modified with bioactive substances [32], or modified with silane preparations in order to achieve better adhesion to the matrix material of the composite [14,15]. In order to improve the strength properties of the composite, in addition to the matrix and diatomite, e.g., polypropylene fibers [7] or multi-walled carbon nanotubes, MWCN [12] is used as the third component. Similarly, in the case of composite on the basis of cement and aluminum powder, natural fibers are added [33].

The composites containing diatomite described in the literature were produced using various techniques and intended for various applications, e.g., for medical purposes [32], agriculture [32], building materials [7,9], for removing heavy metal contaminants [8], or for ensuring the thermal efficiency of the phase change materials PCM [17]. Taking into account the potential use of materials containing diatomite, their various properties were tested. However, there are few literature reports regarding the study of sound absorption properties. It has been shown that in the case of bi-layered diatomite-based foams with a gradient structure and a diatomite content of 21 wt.%, the density of the material is very important. Low-density foams behave like porous materials, i.e., they show a sharp increase in sound absorption at medium frequencies and a high level of sound absorption at high frequencies. High-density foams show a low level of sound absorption in the entire tested range [16].

Diatomite/polyurethane composites were also fabricated, and the effects of diatomite, foam stabilizer, and plasticizer content on the microstructure, compressive strength, and sound absorption properties of diatomite/polyurethane composites were studied. The results show that the porous diatomite/polyurethane composites have good compressive strength and very good sound absorption properties. A diatomite/polyurethane composite containing 65 wt.% diatomite showed excellent sound absorption properties. The sound absorption peak was around 1600 Hz, and the sound absorption coefficient was greater than 0.9. The frequency range whose sound absorption coefficient was greater than 0.56 was more than 2000 Hz [34].

In scientific publications, one can also find research results related to composites based on epoxy resin with the addition of diatomite. One study showed that the composite containing 50 wt.% exhibits an increased elastic modulus of 2.9 MPa, satisfactory sound absorption properties at low frequencies with a modified mean sound absorption (MSAA) of 0.08 (for a sample only 5 mm thick) [35].

Although there are research results confirming the positive effect of diatomite on the acoustic properties of composites, little is known about the effect of diatomite processing and its degree of fragmentation on the sound absorption coefficient. This paper presents a comprehensive and innovative study of the effect of different fractions of diatomite and the effect of its thermal treatment on the sound absorption coefficient of flax/PLA/diatomite composites. To date, no research of this type has been conducted, and how to optimize acoustic parameters by controlling the degree of fragmentation and calcination of diatomites has not been determined. The presented research results make an important contribution to the development of this field of science and may contribute to the wider use of diatomites as materials to support sound absorption. The presented research results are also innovative due to the nature of this material, as the diatomites found in Poland are very different from other such materials, and their effects on sound absorption capabilities are also different. Thanks to the research, it was possible to determine for the first time whether the calcination process of diatomite affects sound absorption capacity, and it was determined whether the degree of fragmentation also has a significant effect. These are very important issues not only from an economic point of view (grinding and calcination costs) but also because of the environmental impact (energy consumption). The results of the presented research are important not only for scientific development but also for the possibility of commercialization in solutions such as panels and sound barriers, vehicle cabin enclosures, etc. The sound absorption capacity of materials used for various types of enclosures, shielding, etc., is important because of the need to protect the health of people working, for example, in the cabins of motor vehicles such as excavators, cranes, drilling rigs, etc. The development of such materials can contribute to the development of solutions to prevent excessive noise in the working environment. In addition, such materials should also be good thermal insulators, which is why this article presents the results of thermal conductivity tests for different variants of diatomite processing.

The conducted research adds to the existing knowledge on the use of diatomites as a filler in composites used for acoustic insulation. So far, the effect of thermal treatment of diatomite on the ability to dampen sound has not been described in the scientific literature, nor has the study of diatomite from the deposit presented in the article, which is quite significant due to the amount of material and its origin.

2. Materials and Methods

The study examined the possibilities of using diatomite grains as a factor in increasing the sound absorption of the material. For this purpose, several variants of the composite material have been developed, differing in terms of type, size, and amount of grain. Two types of diatomite were used, i.e., "non-calcined (raw)-gray" and "calcined-red", three fractions of each type of diatomite differing in grain diameter, i.e., fine ($0 \div 0.063$ mm), medium ($0.5 \div 3$ mm), and coarse ($2 \div 5$ mm). The weight share of diatomite in the composite material was 25 wt.% or 50 wt.%. The percentage of diatomite by weight was selected based on the literature [16,35]. The assumption was to demonstrate the effect of the type of diatomite on the sound absorption of the composite with its share at the level of 25 wt.% and to determine whether and to what extent it will increase at the limiting share of diatomite at the level of 50 wt.%.

The diatomite came from the only producer in Poland (Specialized Mining Company GÓRTECH Sp. z o.o., Cracow, Poland [36]), the diatomite mine in Jawornik Ruski (Podkarpackie Voivodeship). Calcination of the diatomite was carried out at 750 °C using a static method in a laboratory furnace for 24 h. The selection of the calcination temperature was based on the literature [4,37–39] and the authors' previous studies. The temperature was chosen so that it was high enough for the processes of dehydration and dehydroxylation to occur, as well as the removal of organic substances, but not to exceed the temperature of mineral decomposition, and so that liquid phases (sintering) did not occur. Calcined diatomite at 750 °C exhibits optimal properties for damping vibrations due to its unique structural and compositional characteristics that arise from the calcination process. The thermal treatment of diatomite at this temperature enhances its mechanical strength and porosity, which are critical factors in vibration-damping applications. The thermal treatment at 750 °C significantly alters the microstructure of diatomite, leading to an increase in its specific surface area and porosity. This transformation is crucial as a higher porosity allows for better energy dissipation through the material when subjected to vibrational forces. The optimal pore structure formed at this temperature ensures that the material can absorb and dissipate vibrational energy efficiently, making it suitable for applications in construction and materials engineering. The thermal treatment enhances the material's thermal stability and reduces its density, which further contributes to its ability to dampen vibrations effectively. The balance of porosity and strength achieved at this temperature is critical for applications where energy absorption and dissipation are necessary [40–42]. Figure 3 below shows the appearance of the diatomite used in the study.

Figure 3. Diatomite used in the study (fraction 0 ÷ 0.063 mm) (raw, before thermal treatment—left side; calcined diatomite—right side).

The bulk density of the used diatomite in the loose and compacted state was performed according to PN-EN 1097-3:2000 [43]. The tests showed that the raw diatomite in the loose state has a bulk density of 0.82 g/cm^3 (0 ÷ 0.063 mm); 0.85 g/cm^3 (0.5 ÷ 3 mm); 0.68 g/cm^3 (2 ÷ 5 mm), while after the calcination process at 750 °C, its bulk density in the loose state is, respectively, 0.78 g/cm^3; 0.75 g/cm^3; 0.65 g/cm^3. The bulk density in the compacted state is 0.99 g/cm^3; 0.97 g/cm^3; 0.77 g/cm^3 for the non-calcined material and 0.86 g/cm^3; 0.81 g/cm^3; 0.73 g/cm^3 for the calcined material, respectively, for the individual fractions.

Diatomite from the company "GÓRTECH" meets the requirements of radiation hygiene and has a certificate issued by the National Institute of Hygiene (No. HR/B/34/2005). It can be used in buildings with rooms for human habitation or in buildings with livestock enclosures.

Oxide chemical composition analysis was performed for raw diatomite (before thermal treatment) and calcined at 750 °C, as shown in Figure 3. The oxide XRF analysis was performed on a SCHIMADZU EDX-7200 (SHIMADZU Europa GmbH, Duisburg, Germany). The test was carried out in an air atmosphere with holders designed for bulk materials and Mylar film, and the results are shown in Table 1. Only the most important oxides detected in the chemical composition, whose weight share is higher than 0.1, are shown in the table. The oxide composition of the two materials is similar.

Table 1. Oxide analysis for diatomite.

Type of Diatomite	Oxide Composition (wt.%)						
	SiO_2	Al_2O_3	Fe_2O_3	K_2O	SO_3	TiO_2	CaO
Raw diatomite	78.985	15.896	2.623	1.375	0.403	0.350	0.280
Calcinated diatomite	79.594	15.415	2.463	1.402	0.402	0.351	0.280

The mineralogical composition of diatomite was investigated by means of an X-ray diffraction technique using a PANalytical Aeris (Malvern Panalytical, Almelo, The Netherlands) diffractometer. X-ray diffractometry (XRD) is a technique that uses X-ray diffraction to determine the crystalline phases that make up the object under study. Quantitative analysis was performed using the Rietveld method, which was implemented in the HighScore Plus software (version 4.8). Diffractograms were recorded using Cu-Kα radiation in the scan range of 10–100° with a step of 0.003° (2θ) and a time per step of 340 s, using Cu Kα radiation. The X-ray diffraction patterns of raw diatomite and calcined diatomite are presented in Figure 4. Moreover, the results of the quantitative analysis are listed in Table 2. During the study, the weight content of mineral phases, including crystalline silica in the form of quartz, was determined.

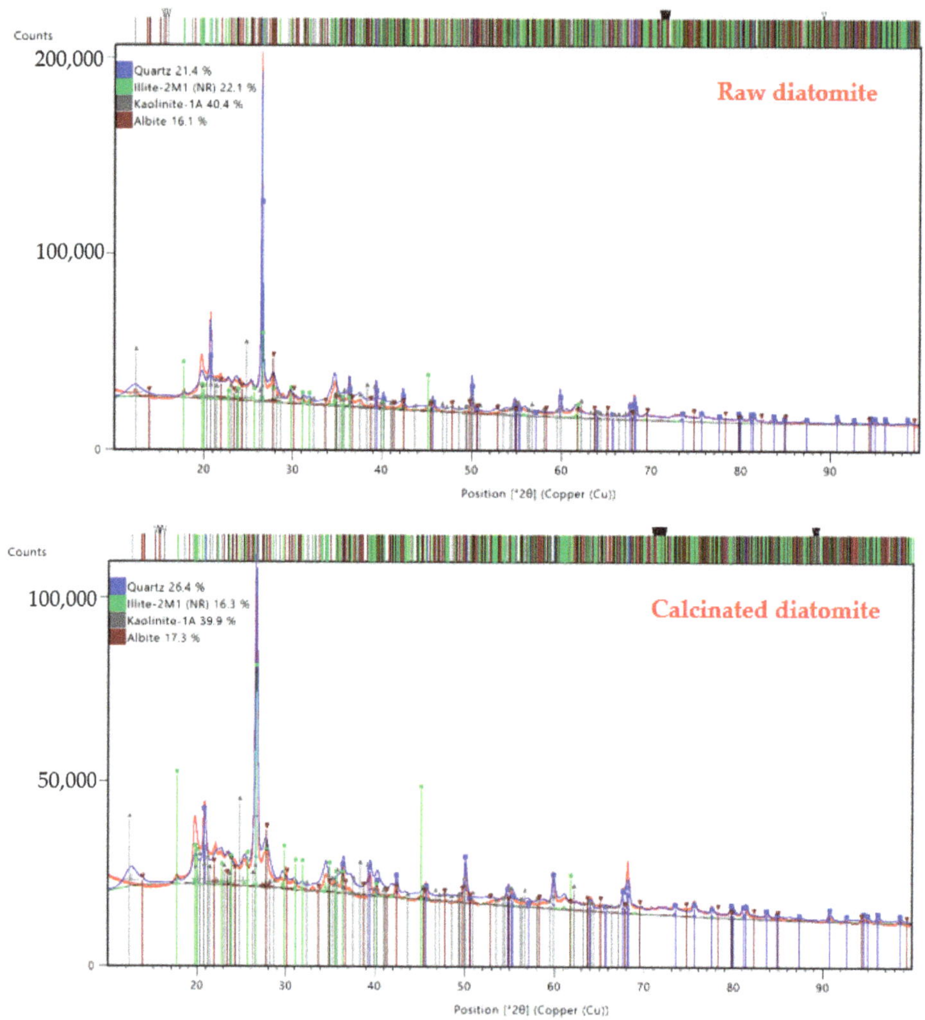

Figure 4. X-ray diffraction patterns of raw diatomite and calcined diatomite.

Table 2. Quantitative analysis of raw diatomite and calcined diatomite.

Type of Diatomite	**Quantitative Share of Phase [%]**			
	Quartz	**Illite**	**Kaolinite**	**Albite**
	01-075-8321	**00-026-0911**	**00-058-2001**	**00-001-0739**
	SiO_2	$Al_2Si_3AlO_{10}(OH)_2$	$Al_2Si_2O_5(OH)_4$	$NaAlSi_3O_8$
Raw diatomite	21.4	22.1	40.4	16.1
Calcined diatomite	26.4	16.3	39.9	17.3

In order to compare the sound absorption of different composite materials, they were manufactured in a way that ensured similar thickness of the test samples and the same external surface exposed to the sound wave. Components enabling biodegradation were used to produce the composite material containing diatomite. Polylactide (PLA) thermoplastic fibers 6.7 dtex/64 mm, under the name of Ingeo Fiber SLN2660D, with a finishing composition containing polylactide resin and no hazardous compounds, supplied by the Far Eastern Textile Ltd., Taiwan [44], were used as the composite matrix material. The melting point of PLA fibers is in the temperature range of 165 ÷ 170 °C. Waste standard flax fibers (flax) from Safilin Sp. z o.o., Poland [45], were used as a reinforcement for the composite.

Both types of fibers were combined in the appropriate proportion (PLA/flax—50 wt.%/50 wt.%) and mixed thoroughly, and then processed into a fleece, from which, after the needling process, a needle-punched nonwoven was created.

Samples of the composite material were produced on a hydraulic press with a water-cooling system from Hydromega Sp. z o. o., Poland. A layer system consisting of layers of nonwoven and layers of diatomite scattered between them was subjected to thermal pressing. In order to ensure a uniform surface of the composite material samples, the upper layer of the system was always made of nonwoven. A multilayer structure was closed in the form of a press, and the heating was turned on. The pressing temperature was 170 °C; the consolidation was carried out under a pressure of 0.58 MPa for 3 min. After the consolidation stage, the composite was cooled to room temperature using the water-cooling system. Each composite was manufactured from the system consisting of 75 wt.% of nonwoven (PLA/flax—50 wt.%/50 wt.%) and 25 wt.% of diatomite or 50 wt.% of nonwoven (PLA/flax—50 wt.%/50 wt.%) and 50 wt.% of diatomite. For a full comparative analysis of the sound absorption of materials containing diatomite, a reference material was also produced, i.e., without the addition of diatomite.

Test samples for testing the sound absorption of composite material, with a thickness of approx. 5 mm were cut with steel dies, i.e., 29 mm diameter circles for high frequencies and 100 mm diameter circles for high frequencies. The thickness of the specimens (5 mm) was chosen based on previous research achievements by other authors [35]. This parameter was selected in order to make the research results comparable to the works of other scientists. Figure 5 shows an example of composites with the addition of diatomite.

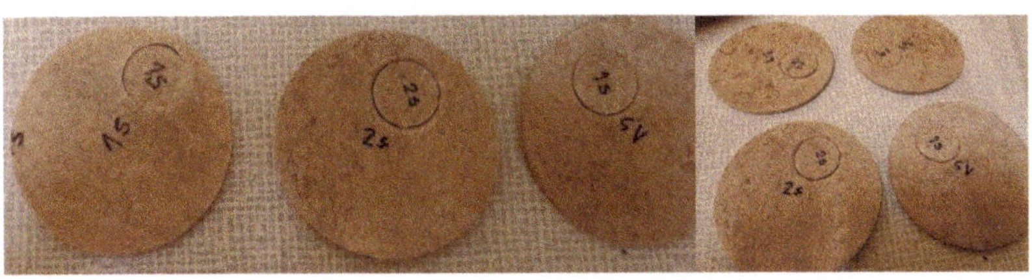

Figure 5. Examples of composites with the addition of diatomite.

3. Experimental

3.1. Morphology Investigations

The morphology of the diatomite was observed using a scanning electron microscope JEOL IT200 (JEOL, Tokyo, Japan). The morphology evaluation focused mainly on identifying the porous structure of diatom carapaces.

3.2. Laser Particle Size Analysis

The particle size characteristics were performed via the particle size analyzer Anton-Paar PSA 1190LD (Anton-Paar, Graz, Austria). Five test measurements were taken for each material, and then the mean and standard deviation were calculated using Kalliope Professional software (version 2.22.1). The tests were conducted only for the 0 ÷ 0.063 mm fraction to obtain an idea of what the size distribution of the finest particles looks like.

3.3. Thermal Conductivity of Diatomite

Measurements of the thermal conductivity of diatomite were carried out on an HFM 446 plate apparatus (Netzsch, Selb, Germany). The instrument operates according to specialized standards such as ASTM C1784 [46], ASTM C518 [47], ISO 8301 [48], EN 12664 [49], and others. The conductivity range is 0.007 to 2.0 W/m × K, the measurement accuracy is ± 1–2%. Temperature regulation and control are verified by a Peltier system. The thermal properties of diatomite were determined using the described device based on the hot and cold plate method. Thermal conductivity was tested in the temperature range of 0–20 °C.

3.4. The Acoustic Properties of the Composite

The acoustic properties of the composite materials were tested using an impedance tube, the so-called Kundt tube, type 4206, in the low-frequency range of normal incidence wave, from 50 Hz to 1600 Hz, and in the high-frequency range from 500 Hz to 6400 Hz. The sound absorption coefficient was determined in accordance with the PN-EN ISO 10534-2 standard [50] based on the following formula. Figure 6 shows the device used for testing to illustrate the test procedure.

$$\alpha = \frac{I_a}{I_i}$$

where:

α—sound absorption coefficient,
I_a—intensity of the sound absorbed,
I_i—intensity of the incident sound.

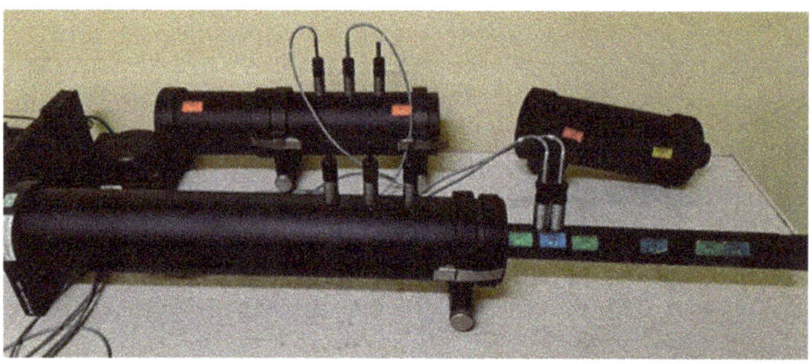

Figure 6. Impedance tube (Brüel & Kjaer, Virum, Denmark).

4. Results

4.1. Morphology Investigations

Figure 7 below shows the morphology of diatomite particles for different fraction sizes without thermal treatment and after thermal treatment. The pictures show different forms of diatomite armor. Various shapes and sizes can be seen, as well as the porosity of these structures. The diatomite armor and its porous structure, which can be seen in the photos, are particularly important in terms of sound attenuation ability. The microstructure of diatomite may play a key role not only in terms of binding diatomite particles to the matrix but also because of its sound absorption ability. SEM images show forms of diatomite—disk-shaped, cylindrical (tubular), and mechanically degraded (crushed) form. It can be seen that after thermal treatment, the appearance of the diatomite particles did not change. Based on SEM observations, it is difficult to say whether there have been significant changes in the morphology of the diatomite particles after calcination. The photos were taken at a magnification of 1000 and 2000×. The marker in the figures corresponds to 10 μm.

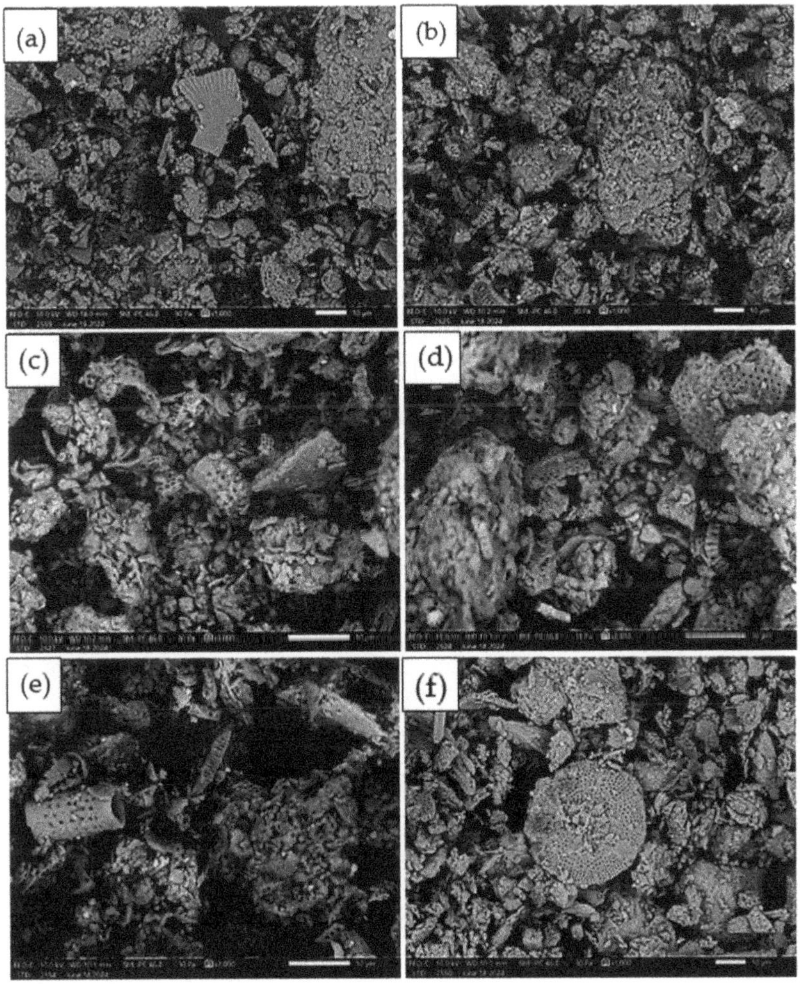

Figure 7. Morphology of diatomite particles: (**a**) 0 ÷ 0.063 mm without thermal treatment; (**b**) 0 ÷ 0.063 mm after thermal treatment; (**c**) 0.5 ÷ 3 mm without thermal treatment; (**d**) 0.5 ÷ 3 mm after thermal treatment; (**e**) 2 ÷ 5 mm without thermal treatment; (**f**) 2 ÷ 5 mm after thermal treatment.

4.2. Laser Particle Size Analysis

Table 3 shows the particle size distribution for the diatomite used in the study. Results for raw material and calcined material are presented. It was found that the calcination process caused a slight increase in the average particle diameter. This may be due to the agglomeration of the finest particles by partial sintering of the low-melting phases present in the diatomite. During the particle size distribution study, ultrasound was used to eliminate the risk of agglomeration of diatomite in aqueous media. The increase in the size of the measured particles after calcination was most likely due to permanent bonding due to thermal treatment and the occurrence of a liquid phase.

Table 3. Particle size distribution of diatomite.

Material	D_{10} [µm]	D_{50} [µm]	D_{90} [µm]	Average Particle Size [µm]	Standard Deviation [µm]
Raw diatomite	2.536	10.633	25.991	13.315	0.398
Calcinated diatomite	3.948	14.488	31.578	17.240	0.075

4.3. Thermal Conductivity of Diatomite

Table 4 shows the results of heat conduction coefficient tests for different fractions of diatomite before and after the calcination process. The results show that the calcination process reduces the thermal conductivity coefficient. The highest values of the heat conduction coefficient were recorded for diatomite with a fraction of 0.5–3 mm and the lowest for 0–0.063 mm. The fraction with the largest particle size had an intermediate thermal coefficient value. For the use of diatomite in sound-absorbing composites that can be used, for example, as covers or fillings in the cabins of heavy construction equipment (vehicles such as excavators, cranes, etc.), thermal insulation is just as important as sound absorption capacity (due to the operation of such equipment in different environmental conditions and at different temperatures).

Table 4. Thermal conductivity of diatomite.

Type of Diatomite	Thermal Conductivity Coefficient λ [W/m × K]
Raw diatomite 0–0.063 mm	0.07725
Calcinated diatomite 0–0.063 mm	0.06969
Raw diatomite 0.5–3 mm	0.15385
Calcinated diatomite 0.5–3 mm	0.12656
Raw diatomite 2–5 mm	0.13260
Calcinated diatomite 2–5 mm	0.11845

4.4. The Acoustic Properties of the Composites

The sound absorption results for the flax/PLA/diatomite composites are presented in the figures below, showing the dependence of the sound absorption coefficient on the sound frequency.

4.4.1. Tests in the Frequency Range of 50–1600 Hz

Figures 8–12 below show a comparison of sound absorption coefficients for composites with different additions of calcined and non-calcined diatomite. The purpose of the study was to determine how the calcination process of diatomite affects the sound absorption capacity of composites with it, as well as how the degree of fineness of the diatomite affects this coefficient.

As can be seen from the tests presented above in Figure 9 for diatomite with a fraction of 0–0.063 mm, the best characteristics in terms of sound absorption coefficient were obtained for the material containing 25 wt.% in non-calcined form. The composite containing 50 wt.% of the same diatomite had similar properties. Samples with calcined diatomite showed a lower sound absorption coefficient. Comparing the graph for the sample unmodified with diatomite, it can be seen that the addition of 25 wt.% of non-calcined diatomite with a particle size of less than 63 µm results in a more than three-fold improvement in acoustic properties.

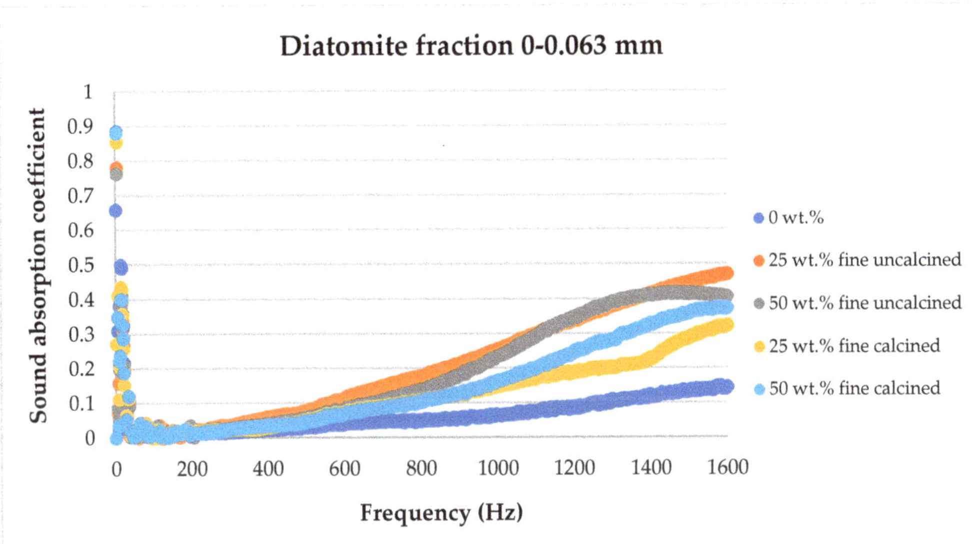

Figure 8. Effect of diatomite addition on the sound absorption coefficient of composites in the low-frequency range—diatomite fraction 0–0.063 mm.

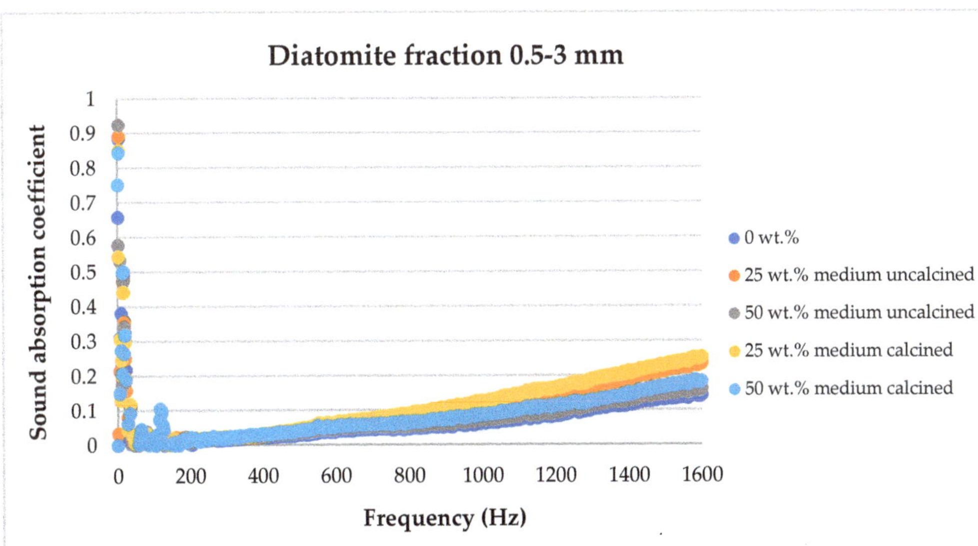

Figure 9. Effect of diatomite addition on the sound absorption coefficient of composites in the low-frequency range—diatomite fraction 0.5–3 mm.

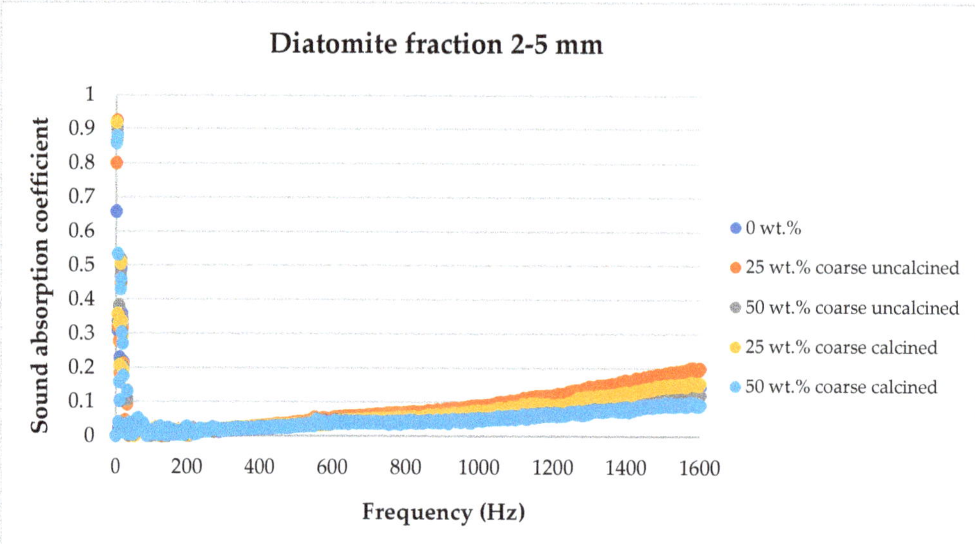

Figure 10. Effect of diatomite addition on the sound absorption coefficient of composites in the low-frequency range—diatomite fraction 2–5 mm.

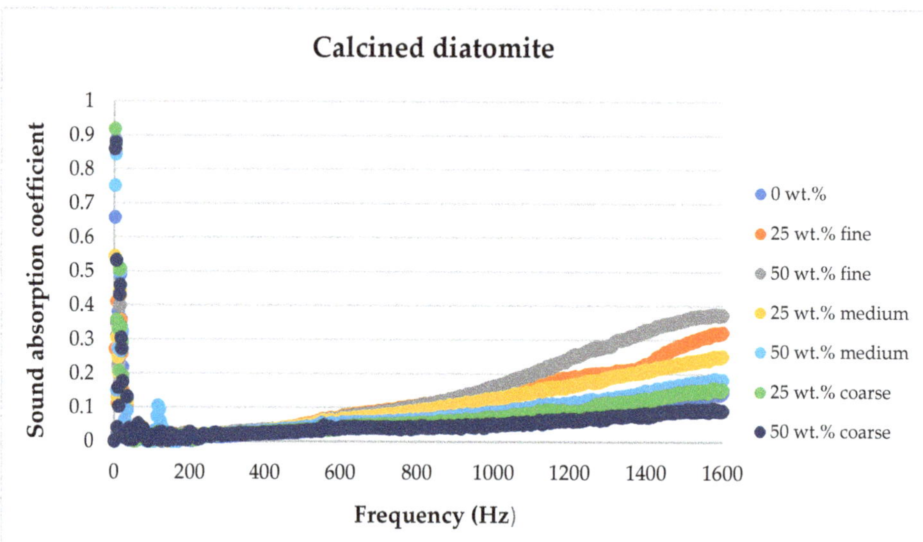

Figure 11. Comparison of the effect of diatomite addition on the sound absorption coefficient of composites in the low-frequency range for calcined diatomite.

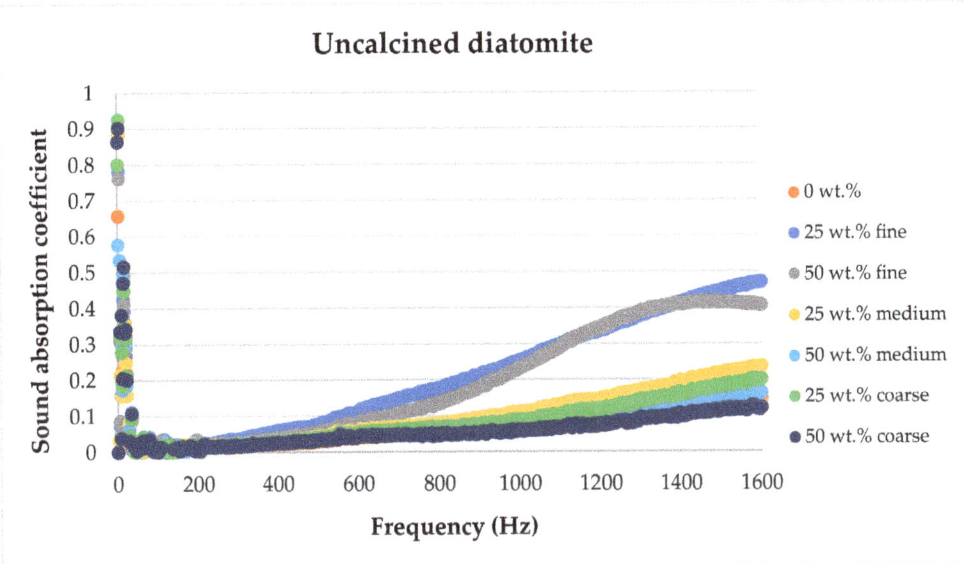

Figure 12. Comparison of the effect of diatomite addition on the sound absorption coefficient of composites in the low-frequency range for uncalcined diatomite.

The tests shown in Figure 10 illustrate the effect of diatomite with particle sizes in the range of 0.5–3 mm on the sound absorption coefficient of the composites. In this case, the most favorable properties were obtained for composites containing 25 wt.% diatomite (both calcined and non-calcined). It is noticeable, however, that compared to the graph presented above (Figure 9) and composites containing diatomite of smaller size, the addition of much larger diatomite particles already has a much less favorable effect than diatomite in dust form.

The addition of diatomite with a particle size of 2–5 mm has little effect on the sound absorption coefficient of the composites. As in the case of smaller fractions, the 25 wt.% diatomite addition also has the most favorable effect.

Comparing different fractions of diatomite and their different shares, for calcined diatomite, it can be concluded that in the frequency range of 50–1600 Hz, the 25 wt.% and 50 wt.% of diatomite addition in the form of 0–0.063 mm fine dust is the most favorable. The use of diatomite with larger fractions affects the sound absorption capacity to a lesser extent and may even negatively affect this parameter. The use of a 50 wt.% diatomite additive with a fraction of 2–5 mm causes a deterioration of the composite's sound attenuation ability compared to the material without the introduction of diatomite.

The use of a non-calcined diatomite additive yielded the best results when the finest fraction below 63 μm was used. The addition of both 50 wt.% and 25 wt.% resulted in a several-fold improvement/increase in sound absorption coefficient.

4.4.2. Tests in the Frequency Range of 500–6400 Hz

Figures 13–17 below present the results of sound absorption coefficient tests in the frequency range of 500–6400 Hz for composites with diatomite in both calcined and non-calcined forms. The three fractions of diatomite were used as before.

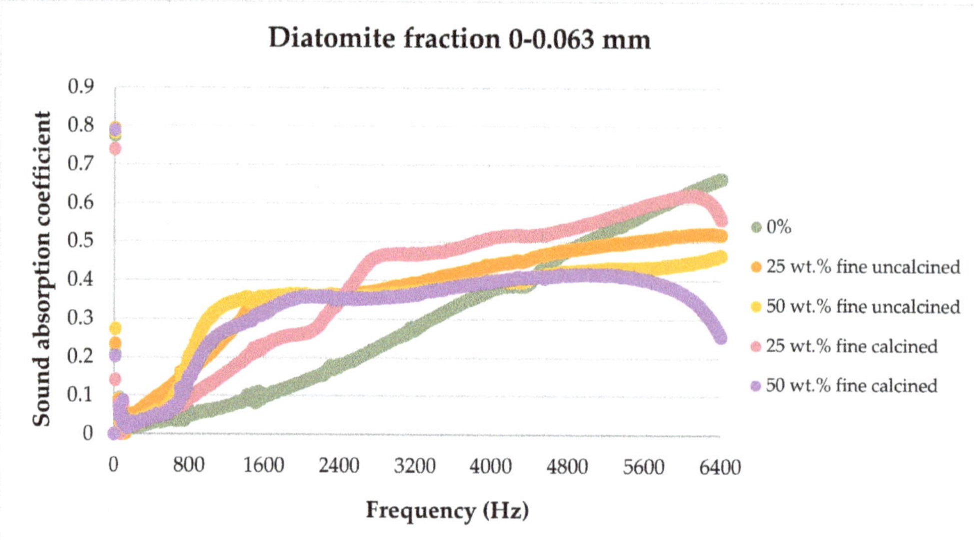

Figure 13. Effect of diatomite addition on the sound absorption coefficient of composites in the high-frequency range—diatomite fraction 0–0.063 mm.

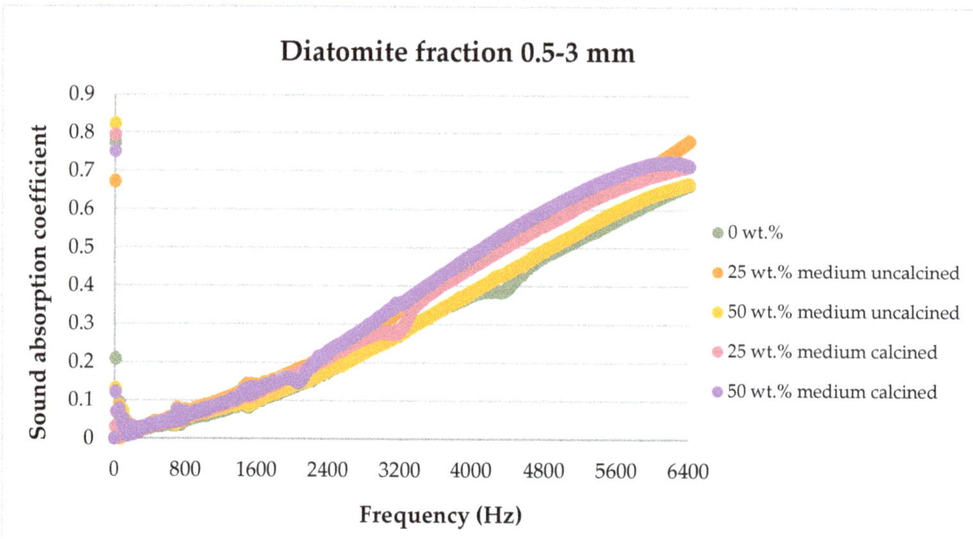

Figure 14. Effect of diatomite addition on the sound absorption coefficient of composites in the high-frequency range—diatomite fraction 0.5–3 mm.

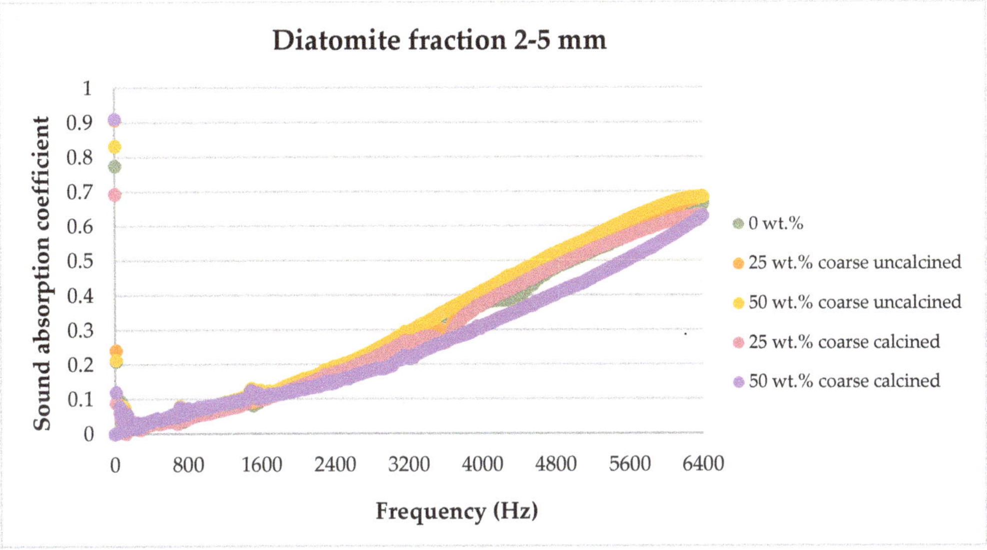

Figure 15. Effect of diatomite addition on the sound absorption coefficient of composites in the high-frequency range—diatomite fraction 2–5 mm.

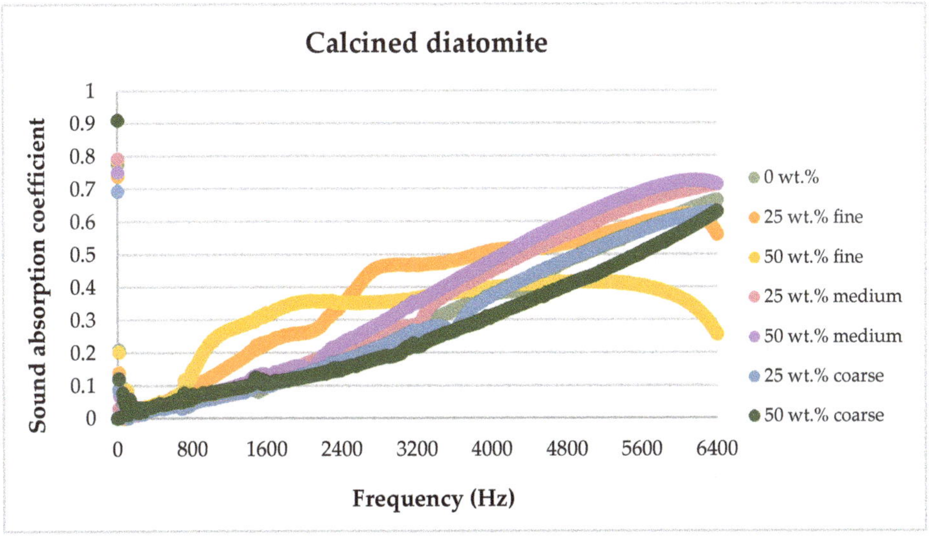

Figure 16. Comparison of the effect of diatomite addition on the sound absorption coefficient of composites in the high-frequency range for calcined diatomite.

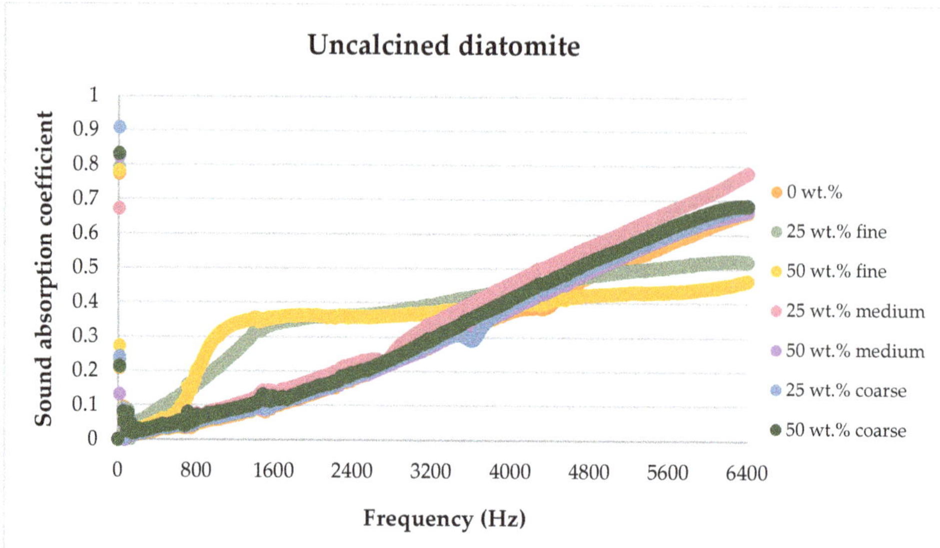

Figure 17. Comparison of the effect of diatomite addition on the sound absorption coefficient of composites in the high-frequency range for uncalcined diatomite.

In the frequency range studied, the addition of the smallest size diatomite did not significantly improve sound absorption capacity. However, it was observed that the addition of both calcined and non-calcined diatomite in the amount of 25 wt.% and 50 wt.% affected the sound absorption capacity in the frequency range from about 800 Hz to about 4500 Hz. Beyond the 5000 Hz frequency range, the addition of diatomite had a negative effect. No significant differences were observed in terms of the effect of heat treatment or the mass proportion of the additive.

The use of larger diatomite fractions also did not yield positive results in the frequency range tested.

Analyzing the effect of diatomite on sound attenuation in the frequency range of 500–6400 Hz, one can see its positive effect in the range of 800–3500 Hz for both the calcined and untreated variants.

5. Discussion

Diatomites and diatomaceous earth are widely used as filter materials, sorbents, carriers for plant protection products and catalysts, thermal insulation, and polishing materials. Recently, the use of diatomite in construction binders has been increasingly investigated [37,38,51], and technologies are being developed to improve their sorption properties in terms of their ability to absorb petroleum substances [4,39]. Diatomites and diatomaceous earth have been used as ceramic fillers in paints and varnishes for many years. The use of diatomite as a filler in anti-corrosion coatings, as well as in other types of coatings or plastics, can bring a number of benefits. According to research carried out by certified masters of radiesthesia, diatomite of Polish origin is a material/additive for neutralizing the negative effects of geopathic (earth) radiation in the B-S-Pc range of the radiesthesia spectrum. This diatomite has a selective effect as it does not suppress the natural bands of radiation in the Cz-UF range that are positive for humans and animals.

It is also known for its antibacterial and fungicidal properties. Research by the authors of this article has shown that diatomite has growth-inhibiting properties against the bacteria tested: Bacillus subtilis, Escherichia coli, and two strains of Staphylococcus aureus. Diatomite showed a linear mycelial growth-limiting effect against the fungus species tested, Aspergillus niger. The results of this research will be presented in a separate scientific article.

The results presented in the paper showed that diatomite from Polish deposits is characterized by SiO_2 content not exceeding 80 wt.%. The results of these studies are consistent with the work of other scientists [5,38]. Analysis of mineral phases revealed the presence of crystalline phases such as quartz, illite, kaolinite, and albite in the diatomite, which also coincides with the findings of other researchers [39,52,53]. Studies of diatomite morphology have shown that diatomite has a very porous structure, a highly desirable effect in terms of acoustic capabilities [54–56]. A particle size analysis of diatomite powder showed that the calcination process caused an increase in the particle size of the material. This increase was most likely due to permanent binding due to thermal treatment and the presence of a liquid phase [4,38,51]. The thermal treatment of diatomite also caused a decrease in the thermal conductivity coefficient for each of the fractions tested. The smallest diatomite particles obtained the lowest thermal conductivity coefficient after calcination [57]. The calcination process has a positive effect on lowering the thermal conductivity coefficient.

The acoustic properties of diatomite composites are significantly influenced by both thermal conductivity and porosity. Diatomite, a naturally occurring siliceous sedimentary rock, is characterized by its high porosity and low thermal conductivity, which are critical factors in determining its acoustic behavior. The unique microstructure of diatomite, composed of numerous interconnected pores, contributes to its sound absorption capabilities. As porosity increases, the ability of the material to attenuate sound waves also improves, leading to enhanced acoustic performance [40,58]. The relationship between porosity and acoustic properties can be attributed to the mechanisms of sound wave propagation through porous materials. Increased porosity generally results in a greater surface area for sound wave interaction, which can lead to higher sound absorption coefficients. For instance, studies have shown that diatomite composites with porosities exceeding 90% exhibit notable sound insulation properties, as the air trapped within the pores acts as a damping medium for sound waves [59]. Furthermore, the acoustic performance is also affected by the size and distribution of the pores; larger and more uniformly distributed pores tend to enhance sound absorption [35]. On the other hand, thermal conductivity plays a crucial role in the thermal management of diatomite composites, which can indirectly influence their acoustic properties. The thermal conductivity of diatomite is relatively low, which is beneficial for applications requiring thermal insulation. However, when diatomite is incorporated into composites with materials that have higher thermal conductivity, such as expanded graphite or carbon nanotubes, the overall thermal conductivity of the composite can be significantly enhanced [60,61]. This modification can lead to changes in the microstructural interactions within the composite, potentially affecting its acoustic properties. For example, the incorporation of conductive fillers can alter the density and stiffness of the composite, which may influence how sound waves propagate through the material [62]. Additionally, the interplay between porosity and thermal conductivity can create a complex relationship affecting the acoustic performance of diatomite composites. High porosity typically correlates with lower thermal conductivity, which can enhance sound absorption. However, if the thermal conductivity is increased through the addition of conductive materials, it may lead to a denser structure that could diminish the sound absorption capabilities [35,59]. Therefore, optimizing the balance between porosity and thermal conductivity is essential for achieving desired acoustic properties in diatomite composites. In conclusion, the effects of thermal conductivity and porosity growth on the acoustic properties of diatomite composites are interrelated. Increased porosity generally enhances sound absorption, while thermal conductivity influences the microstructural characteristics of the composite. The careful selection and combination of materials can lead to improved acoustic performance, making diatomite composites suitable for various applications in sound insulation and thermal management.

A number of these properties make diatomites an extremely attractive material that can be used as a filler or active additive in various types of plastics. The results presented in this article show that properly prepared diatomite can contribute to an increase in the sound attenuation coefficient and can be used in a variety of applications.

The studies conducted and presented above indicate that the calcination process is not valid for the use of diatomite in PLA-based composites to achieve better sound attenuation performance. The calcination of diatomite is most relevant for its use as a sorbent for petroleum substances. For other applications, the calcination process is not as important and may not be economically or environmentally viable. The studies presented above have also shown that the most beneficial way to improve sound attenuation capacity is to use diatomite with the smallest possible fraction.

The results obtained for the tested fiber composites with the addition of diatomite confirmed literature reports on similar composites with the addition of cellulose sub-microfibers or cellulose ultra-short and ultra-fine fibers. As the literature indicates [63,64], the content of sub-microfibers or ultra-short and ultra-fine fibers with a diameter of approx. 11 μm, which are thinner than the standard fibers used as reinforcement material in the composite, results in an adequately increased sound absorption coefficient of the composites. This effect is greater if the standard fibers are replaced by thinner fibers, in which case there is a larger fiber surface area involved in the interaction with the sound wave. Using diatomite particles of the smallest possible size, i.e., smaller than the transverse dimension of the standard fibers (the length of the elementary flax fiber varies on average between 1 mm and 120 mm, and the diameter also varies and ranges from 9.6 μm to 201.6 μm), the effect is similar. In our study, the finest fraction of diatomite, i.e., 0–0.063 mm (up to 63 μm), raw, i.e., with a smaller average particle size than after calcination, proved to be the best fraction. As indicated by the results of the study (Table 3), the average particle size for raw diatomite is 13.315 μm, while for calcinated diatomite, it is 17.240 μm.

6. Conclusions

Several conclusions can be drawn from the above discussion of the research results to conclude the study work as follows:

I. Diatomite from Polish deposits is characterized by SiO_2 content not exceeding 80 wt.%. In diatomite, there are crystalline phases, such as quartz, illite, kaolinite, and albite;

II. The thermal treatment caused an increase in the particle size of the material. This increase was most likely due to permanent binding due to thermal treatment and the presence of a liquid phase;

III. The smallest diatomite particles obtained the lowest thermal conductivity coefficient after calcination. The calcination process lowers the thermal conductivity coefficient. When diatomite is used in sound-absorbing composites, thermal insulation is as important as sound absorption capacity (due to the operation of such equipment in different environmental conditions and at different temperatures);

IV. The smallest fraction of diatomite particles and variants without thermal treatment give the best effect in terms of sound absorption. The calcination process does not positively affect the improvement of acoustic properties.

The most important research finding in this study is that the calcination process does not positively affect acoustic properties. So far, the thermal treatment of diatomite has been a very desirable process, as it significantly improved the sorption capacity of the material. For acoustic insulation, calcination is not necessary. This effect is very surprising and has a very positive impact on both environmental and financial aspects. Thermal treatment involves additional costs of electricity consumption, which can be avoided when producing diatomite composites for acoustic applications.

Different fractions of diatomite from the only mine of this raw material in Poland were tested. Diatomite was used as a filler in PLA-based composites with the addition of fibers, and the effect of diatomite on the sound-damping capacity of such prepared composites was studied. Heat-treated calcined and untreated material was used, and the following proportions of diatomite were used: 0 wt.%, 25 wt.%, and 50 wt.%. As a result of the tests carried out, positive preliminary results were obtained on the sound-absorbing properties of the composites with the addition of diatomite. Some variants of the produced composites, as well as the results of the obtained tests, can be useful in the design of vibro-acoustic protection, such as sound-absorbing enclosures, building partitions, acoustic screens, and others.

Author Contributions: Conceptualization, A.P., E.G. and M.Ł.; methodology, A.P., E.G. and M.Ł.; validation, A.P. and M.Ł.; formal analysis, A.P., E.G. and M.Ł.; investigation, A.P. and E.G.; resources, M.Ł. and E.G.; data curation, A.P.; writing—original draft preparation, A.P., E.G. and M.Ł.; writing—review and editing, A.P. and M.Ł.; visualization, A.P. and M.Ł.; supervision, M.Ł. and K.S.; funding acquisition, M.Ł. and K.S. All authors have read and agreed to the published version of the manuscript.

Funding: This work has been partially financed by the National Centre for Research and Development in Poland under the grant: "Development and demonstration of technologies for the production of highly effective diatomite-based sorbents and diatomite fillers" realized within the project 1/4.1.4/2020 funded by the National Centre for Research and Development, Project No.: POIR.04.01.04-00-0032/20.

Institutional Review Board Statement: Not applicable.

Informed Consent Statement: Not applicable.

Data Availability Statement: The original contributions presented in the study are included in the article, further inquiries can be directed to the corresponding authors.

Acknowledgments: Special thanks to Agnieszka Przybek, project manager of Pearls of Science. Part of the research presented in the paper was carried out within the framework of project PN/01/0067/2022, funded by the Ministry of Science and Higher Education in Poland.

Conflicts of Interest: Author Krzysztof Smoroń was employed by the company Specialized Mining Company GÓRTECH Sp. z o.o. The remaining authors declare that the research was conducted in the absence of any commercial or financial relationships that could be construed as a potential conflict of interest.

References

1. Bońda, R. Diatomite rock. In *Balance of Resources of Fossil Deposits in Poland as of December 31, 2023*; Szuflicki, M., Malon, A., Tymiński, M., Eds.; National Geological Institute—National Research Institute: Warsaw, Poland, 2024; pp. 80–81.
2. Crangle, R.D. Mineral Commodity Summaries 2024. U.S. Geological Survey. 2024. Available online: https://pubs.usgs.gov/periodicals/mcs2024/mcs2024-diatomite.pdf (accessed on 17 July 2024).
3. Where Is Diatomite Found. Available online: https://www.mapsofworld.com/minerals/world-diatomite-producers.html (accessed on 17 July 2024).
4. Łach, M.; Pławecka, K.; Marczyk, J.; Ziejewska, C.; Hebdowska-Krupa, M.; Nykiel, M.; Hebda, M.; Miernik, K.; Mierzwiński, D.; Korniejenko, K.; et al. Use of diatomite from Polish fields in sustainable development as a sorbent for petroleum substances. *J. Clean. Prod.* **2023**, *389*, 136100. [CrossRef]
5. Lutyński, M.; Sakiewicz, P.; Lutyńska, S. Characterization of Diatomaceous Earth and Halloysite Resources of Poland. *Minerals* **2019**, *9*, 670. [CrossRef]
6. Figarska-Warchoł, B.; Stańczak, G.; Rembiś, M.; Toboła, T. Diatomaceous rocks of the Jawornik deposit (the Polish Outer Carpathians)—Petrophysical and petrographical evaluation. *Geol. Geophys. Environ.* **2015**, *41*, 311–331. [CrossRef]
7. Genclel, O.; del Coz Diaz, J.J.; Sutcu, M.; Koksal, F.; Rabanal, F.P.A.; Martínez-Barrera, G. A novel lightweight gypsum composite with diatomite and polypropylene fibers. *Constr. Build. Mater.* **2016**, *113*, 732–740. [CrossRef]
8. Dai, S.; Zhang, L.; Hao, C.; Qi, P.; Wang, G.; Ma, W. Preparation and characterization of zeolite/diatomite composite materials. *Mater. Res. Express* **2019**, *6*, 125516. [CrossRef]
9. Pavlíková, M.; Záleská, M.; Pivák, A.; Jankovský, O.; Lauermannová, A.-M.; Lojka, M.; Antoncík, F.; Pavlík, Z. MOC-diatomite composites filled with multi-walled carbon nanotubes. *Materials* **2021**, *14*, 4576. [CrossRef]
10. Rubino, C.; Lama, G.C.; Liuzzi, S.; Martellotta, F.; Liguori, B.; Recupido, F.; Verdolotti, L.; Sorrentino, L. Tailoring porosity and acoustic properties in bi-layered diatomite-based foams through multiscale structural approach. *Constr. Build. Mater.* **2024**, *430*, 136480. [CrossRef]

11. Wei, H.; He, Q.; Jiao, Y.; Chen, J.; Hu, M. Evaluation of anti-icing performance for crumb rubber and diatomite compound modified asphalt mixture. *Constr. Build. Mater.* **2016**, *107*, 109–116. [CrossRef]
12. Wu, W.L.; Chen, Z. Modified-diatomite reinforced rubbers. *Mater. Lett.* **2017**, *209*, 159–162. [CrossRef]
13. Wu, W.; Cong, S. Silica and diatomite fillers modified fluorine rubber composites treated by silane-coupling agents. *J. Vinyl Addit. Technol.* **2020**, *26*, 55–61. [CrossRef]
14. Zhang, C.; Wang, J.; Song, S. Preparation of a novel type of flame retardant diatomite and its application in silicone rubber composites. *Adv. Powder Technol.* **2019**, *30*, 1567–1575. [CrossRef]
15. Qian, T.; Li, J.; Deng, Y. Pore structure modified diatomite supported PEG composites for thermal energy storage. *Sci. Rep.* **2016**, *6*, 32392. [CrossRef]
16. Karaman, S.; Karaipekli, A.; Sari, A.; Bicer, A. Polyethylene glycol (PEG)/diatomite composite as a novel form-stable phase change material for thermal energy storage. *Sol. Energy Mater. Sol. Cells* **2011**, *95*, 1647–1653. [CrossRef]
17. Xu, T.; Wu, F.; Zou, T.; Li, J.; Yang, J.; Zhou, X.; Liu, D.; Bie, Y. Development of diatomite-based shape-stabilized composite phase change material for use in floor radiant heating. *J. Mol. Liq.* **2022**, *348*, 118372. [CrossRef]
18. Song, J.; Cao, X.; Huang, Z. Diatomite-chitosan composite with abundant functional groups as efficient adsorbent for vanadium removal: Key influencing factors and influence of surface functional groups. *J. Mol. Liq.* **2022**, *367 Pt A*, 120428. [CrossRef]
19. Fu, Y.; Xu, X.; Huang, Y.; Hu, Y.; Chen, Q.; Wu, Y. Preparation of new diatomite–chitosan composite materials and heir adsorption properties and mechanism of Hg(II). *R. Soc. Open Sci.* **2017**, *4*, 170829. [CrossRef]
20. Dobrosielska, M.; Dobrucka, R.; Brzakalski, D.; Kozera, P.; Martyła, A.; Gabriel, E.; Kurzydłowski, K.J.; Przekop, R. Polyamide 11 composites reinforced with diatomite biofiller—Mechanical, rheological and crystallization properties. *Polymers* **2023**, *15*, 1563. [CrossRef] [PubMed]
21. Kucuk, F.; Sismanoglu, S.; Kanbur, Y.; Tayfun, U. Effect of silane-modification of diatomite on its composites with thermoplastic polyurethane. *Mater. Chem. Phys.* **2020**, *256*, 123683. [CrossRef]
22. Cacciotti, I.; Rinaldi, M.; Fabbrizi, J.; Nanni, F. Innovative polyetherimide and diatomite based composites: Influence of the diatomite kind and treatment. *J. Mater. Res. Technol.* **2019**, *8*, 1737–1745. [CrossRef]
23. Liang, J.Z. Effects of extrusion conditions on die-swell behavior of polypropylene/ diatomite composite melts. *Polym. Test.* **2008**, *27*, 936–940. [CrossRef]
24. Cacciotti, I.; Mori, S.; Cherubini, V.; Nanni, F. Eco-sustainable systems based on poly (lactic acid), diatomite and coffee grounds extract for food packaging. *Int. J. Biol. Macromol.* **2018**, *112*, 567–575. [CrossRef] [PubMed]
25. Lamastra, F.R.; De Angelis, R.; Antonucci, A.; Salvatori, D.; Prosposito, P.; Casalboni, M.; Congestri, R.; Melino, S.; Nanni, F. Polymer composite random lasers based on diatom frustules as scatterers. *RSC Adv.* **2014**, *4*, 61809–61816. [CrossRef]
26. Fazli, Y.; Khezri, K. Mesoporous diatomite-filled PMMA by in situ reverse atom transfer radical polymerization. *Colloid Polym. Sci.* **2017**, *29*, 247–257. [CrossRef]
27. Lv, J.; Sun, B.; Jin, J.; Jiang, W. Mechanical and slow-released property of poly (acrylamide) hydrogel reinforced by diatomite. *Mater. Sci. Eng. C* **2019**, *99*, 315–321. [CrossRef]
28. Li, X.; Li, X.; Wang, G. Surface modification of diatomite using polyaniline. *Mater. Chem. Phys.* **2007**, *102*, 140–143. [CrossRef]
29. Cavodeau, F.; Otazaghine, B.; Sonnier, R.; Lopez-Cuesta, J.M.; Delaite, C. Fire retardancy of ethylene-vinyl acetate composites—Evaluation of synergistic effects between ATH and diatomite fillers. *Polym. Degrad. Stab.* **2016**, *129*, 246–259. [CrossRef]
30. Gültürk, E.A.; Güden, M.; Taşdemirci, A. Calcined and natural frustules filled epoxy matrices: The effect of volume fraction on the tensile and compression behavior. *Compos. Part B Eng.* **2013**, *44*, 491–500. [CrossRef]
31. Taşdemirci, A.; Yüksel, S.; Karsu, D.; Gültürk, E.; Hall, I.W.; Güden, M. Diatom frustule-filled epoxy: Experimental and numerical study of the quasi-static and high strain rate compression behavior. *Mater. Sci. Eng. A* **2008**, *480*, 373–382. [CrossRef]
32. Kabieva, S.; Zhumanazarova, G.; Zhaslan, R.; Zhumabayeva, G.; Ukhov, A.; Fedorishin, D.; Gubankov, A.; Tarikhov, F.; Yerkhan, O.; Kurzina, I.; et al. Obtaining new biocompatible composite materials with antibacterial properties based on diatomite and biologically active compounds. *Molecules* **2024**, *29*, 1608. [CrossRef]
33. Yoon, J.; Kim, H.; Koh, T.; Pyo, S. Microstructural characteristics of sound absorbable porous cement-based materials by incorporating natural fibers and aluminum powder. *Constr. Build. Mater.* **2020**, *243*, 118167. [CrossRef]
34. Lin, J.; Guo, J.; Zhao, Y.; Duan, L.; Jin, S. Microstructure and sound absorption property of diatomite/polyurethane porous composites. *Fuhe Cailiao Xuebao/Acta Mater. Compos. Sin.* **2014**, *31*, 1476–1480.
35. Nguyen, Q.-B.; Vahabi, H.; Rios de Anda, A.; Versace, D.-L.; Langlois, V.; Perrot, C.; Nguyen, V.-H.; Naili, S.; Renard, E. Dual UV-Thermal Curing of Biobased Resorcinol Epoxy Resin-Diatomite Composites with Improved Acoustic Performance and Attractive Flame Retardancy Behavior. *Sustain. Chem.* **2021**, *2*, 24–48. [CrossRef]
36. DIATO Diatomite Sorbent. Available online: https://diato.pl (accessed on 17 July 2024).
37. Brudny, K.; Łach, M.; Bąk, A.; Pławecka, K.; Korniejenko, K. Application of diatomite as a substitute for fly ash in foamed geopolymers. *J. Phys. Conf. Ser.* **2023**, *2423*, 012028. [CrossRef]
38. Pławecka, K.; Bąk, A.; Hebdowska-Krupa, M.; Łach, M. The Use of Calcined Diatomite as an Additive to Geopolymeric Materials. *Mater. Proc.* **2023**, *13*, 28. [CrossRef]
39. Marczyk, J.; Pławecka, K.; Hebdowska-Krupa, M.; Nykiel, M.; Łach, M. Research on diatomite from Polish deposits and the possibilities of its use. *J. Achiev. Mater. Manuf. Eng.* **2022**, *115*, 5–15. [CrossRef]

40. Figarska-Warchoł, B.; Rembiś, M.; Stańczak, G. The impact of calcination on changes in the physical and mechanical properties of the diatomites of the Leszczawka Member (the Outer Carpathians, Poland). *Geol. Geophys. Environ.* **2019**, *45*, 269. [CrossRef]
41. Ediz, N.; Bentli, I.; Tatar, I. Improvement in filtration characteristics of diatomite by calcination. *Int. J. Miner. Process.* **2010**, *94*, 129–134. [CrossRef]
42. Hamed, A. A brief and general overview on diatoms and their applications. *Egypt. J. Phycol.* **2023**, *24*, 1–53. [CrossRef]
43. *PN-EN 1097-3:2000*; Tests of Mechanical and Physical Properties of Aggregates—Determination of Bulk Density and Coarseness. Polish Committee for Standardization: Warsaw, Poland, 2000.
44. Far Eastern Group. Available online: https://www.feg.com.tw/en/home/index.aspx (accessed on 17 July 2024).
45. Safilin. Available online: https://www.safilin.fr/?lang=pl (accessed on 17 July 2024).
46. *ASTM C1784*; Standard Test Method for Using a Heat Flow Meter Apparatus for Measuring Thermal Storage Properties of Phase Change Materials and Products. International American Society for Testing of Materials: West Conshohocken, PA, USA, 2020.
47. *ASTM C518*; Standard Test Method for Steady-State Thermal Transmission Properties by Means of the Heat Flow Meter Apparatus. International American Society for Testing of Materials: West Conshohocken, PA, USA, 2021.
48. *ISO 8301*; Thermal Insulation—Determination of Steady-State Thermal Resistance and Related Properties—Heat Flow Meter Apparatus. International Standards Organization: London, UK, 1991.
49. *EN 12664*; Thermal Performance of Building Materials and Products. Determination of Thermal Resistance by Means of Guarded Hot Plate and Heat Flow Meter Methods. Dry and Moist Products of Medium and Low Thermal Resistance. European Committee for Standardisation: Brussels, Belgium, 2001.
50. *PN-EN ISO 10534-2:2024-04*; Acoustics—Determination of acoustic properties in impedance tubes—Part 2: Two-Microphone Method for Perpendicular Sound Absorption Coefficient and Perpendicular Sound Insulation. Polish Committee for Standardization: Warsaw, Poland, 2024.
51. Nykiel, M.; Korniejenko, K.; Setlak, K.; Melnychuk, M.; Polivoda, N.; Kozub, B.; Hebdowska-Krupa, M.; Łach, M. The Influence of Diatomite Addition on the Properties of Geopolymers Based on Fly Ash and Metakaolin. *Materials* **2024**, *17*, 2399. [CrossRef]
52. Stempkowska, A.; Gawenda, T.; Smoroń, K. The Diatomite Grinding Technology Concept for the Protection of Diatomite Shells and the Control of Product Grading. *Materials* **2024**, *17*, 3662. [CrossRef]
53. Kobylinska, N.; Dudarko, O.; Gładysz-Płaska, A.; Tertykh, V.A.; Majdan, M. Optimal Synthesis of Novel Phosphonic Acid Modified Diatomite Adsorbents for Effective Removal of Uranium(VI) Ions from Aqueous Solutions. *Materials* **2023**, *16*, 5263. [CrossRef] [PubMed]
54. Hoang, A.T.; Nižetić, S.; Duong, X.Q.; Rowinski, L.; Nguyen, X.P. Advanced super-hydrophobic polymer-based porous absorbents for the treatment of oil-polluted water. *Chemosphere* **2021**, *277*, 130274. [CrossRef] [PubMed]
55. Du, W.; Dai, G.; Wang, B.; Li, Z.; Li, L. Biodegradable porous organosilicone-modified collagen fiber matrix: Synthesis and high oil absorbency. *J. Appl. Polym. Sci.* **2018**, *135*, 46264. [CrossRef]
56. Yadav, S.; Illa, M.P.; Rastogi, T.; Sharma, C.S. High absorbency cellulose acetate electrospun nanofibers for feminine hygiene application. *Appl. Mater. Today* **2016**, *4*, 62–70. [CrossRef]
57. Ivanov, S.E.; Belyakov, A.V. Diatomite and its application. *Glass Ceram.* **2008**, *65*, 1–4. [CrossRef]
58. Skvortsov, A.A.; Luk'yanov, M.N.; Novitsan, Y.V. Sintering and Mechanical Properties of High-Porosity Ceramics Based on Diatomite. *Solid State Phenom.* **2017**, *269*, 71–77. [CrossRef]
59. Skvortsov, A.A.; Nikolaev, V.K.; Luk'yanov, M.N.; Chebeneva, I.E. On the issue of deformation and destruction of porous ceramics based on diatomite. *Phys. Solid State* **2023**, *1*, 120. [CrossRef]
60. Qian, T.; Li, J.; Min, X.; Guan, W.; Deng, Y.; Ning, L. Enhanced thermal conductivity of PEG/diatomite shape-stabilized phase change materials with Ag nanoparticles for thermal energy storage. *J. Mater. Chem. A* **2015**, *3*, 8526–8536. [CrossRef]
61. Qian, T.; Li, J.; Ma, H.; Yang, I. Adjustable thermal property of polyethylene glycol/diatomite shape-stabilized composite phase change material. *Polym. Compos.* **2016**, *37*, 854–860. [CrossRef]
62. Tang, F.; Su, D.; Tang, Y.; Fang, G. Synthesis and thermal properties of fatty acid eutectics and diatomite composites as shape-stabilized phase change materials with enhanced thermal conductivity. *Sol. Energy Mater. Sol. Cells* **2015**, *141*, 218–224. [CrossRef]
63. Gliścińska, E.; Michalak, M.; Krucińska, I.; Kazimierczak, J.; Bloda, A.; Ciechańska, D. Sound Absorbing Composites from Nonwoven and Cellulose Submicrofibres. *J. Chem. Chem. Eng.* **2013**, *7*, 942–948.
64. Krucińska, I.; Gliścińska, E.; Michalak, M.; Ciechańska, D.; Kazimierczak, J.; Bloda, A. Sound-absorbing green composites based on cellulose ultra-short/ultra-fine fibres. *Text. Res. J.* **2015**, *85*, 646–657. [CrossRef]

MDPI AG
Grosspeteranlage 5
4052 Basel
Switzerland
Tel.: +41 61 683 77 34

Materials Editorial Office
E-mail: materials@mdpi.com
www.mdpi.com/journal/materials

www.ingramcontent.com/pod-product-compliance
Lightning Source LLC
Chambersburg PA
CBHW040716270926
51850CB00023B/2651
9783725871582